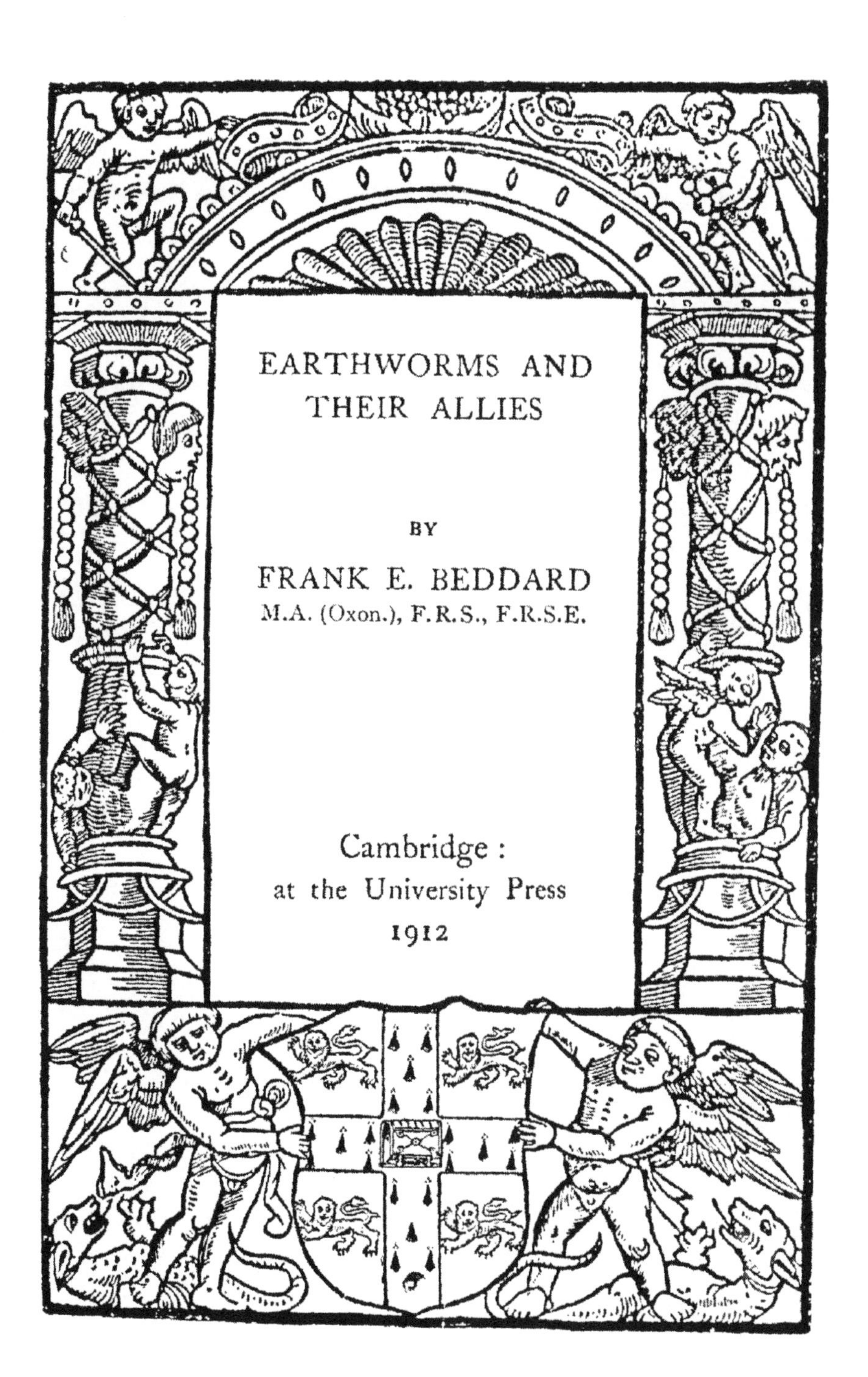

EARTHWORMS AND THEIR ALLIES

BY

FRANK E. BEDDARD
M.A. (Oxon.), F.R.S., F.R.S.E.

Cambridge :
at the University Press
1912

PREFACE

THE importance of earthworms in questions relating to geographical distribution is so universally admitted that it seemed to me convenient to embody in a short volume the principal facts.

It became necessary in order to accomplish this task in an adequate fashion to preface the distributional facts with some anatomical and zoological data. I have reduced this section of the book to a minimum and I trust that the illustrations will enable the reader, who is not specially acquainted with the structure of these animals, to obtain an idea of their general features and variability in external character and internal anatomy. While the use of technical terms is inevitable in presenting such details, it will be found, I think, that a reference to the figures will render them intelligible.

Since this volume mainly deals with the phenomena of distribution, I have included in my survey nearly all of the usually admitted genera of worms, particularly of the terrestrial forms, which are in the light of our present knowledge the more important in considering this subject.

F. E. B.

ZOOLOGICAL SOCIETY OF LONDON.
December, 1911

CONTENTS

CHAPTER I

STRUCTURAL AND SYSTEMATIC

THE group of segmented, bristle-bearing, worms, termed Oligochaeta by zoologists, comprises what are popularly known as earthworms together with certain forms, inhabitants of ponds, lakes and rivers, which are not so well known as to have received a more distinctive name than merely 'worms.' Their next allies are apparently the leeches and—a little more remote—marine bristle-bearing worms termed Polychaeta; the three groups, together with perhaps a certain number of other forms belonging to smaller groups, constitute the Annelida which are a distinct and separate assemblage of invertebrate animals.

The most interesting features about these Oligochaetous worms are their very great anatomical variation and the facts of their distribution over the globe. Their importance as geological agents in levelling the ground was made known a long time ago by Darwin, and that aspect of earthworms has remained in much the same position as Darwin left it. We shall concern ourselves here only with the structure, habits, and range of the earthworms and their immediate allies, the aquatic Oligochaeta. These three aspects of the animals dovetail into each other more thoroughly than is the case with some other groups. This is due to the fact that they have of late years been very thoroughly studied from the anatomical and distributional side. So lately as 1889, M. Vaillant in a very comprehensive treatise was only able to enumerate 369 species, of which a large number were but incompletely differentiated, and some are no longer admitted. There are at the moment of writing perhaps 1500 species, the vast majority of which are well known owing to careful investigation. Furthermore there are but few parts of the world, and these are not of large area, from which earthworms at any rate have not been gathered. Though there can be no doubt that a very considerable number of species await discovery, it would seem that we are in possession of information which is not likely to be seriously affected by future researches.

THE ANATOMY OF EARTHWORMS.

Although it is not contemplated to make the present volume a guide to the structure of this group of worms, it is necessary to give some little anatomical sketch of the group in order first of all to illustrate their diversity of structure, secondly to

give reasons for the classification of them, and thirdly to enable the reader to realise certain structural details which it is absolutely necessary to give some account of in order to explain other matters.

It is for example impossible to attempt any account of the fitness of some of these animals for their terrestrial life and of others for an aquatic life without treating of anatomy to some extent.

I shall take one particular species as a type and indicate later the principal divergencies shown by other forms. According to the general opinion among those who have studied the Oligochaeta I take as a representative form a Megascolecid (this and the other families are dealt with *seriatim* on p. 14 et seq.), as this group is presumed to be the oldest, and within that group a representative of the genus *Notiodrilus* which is with some reason held to be the most primitive genus in the group. Finally I have no particular reason for selecting the species *Notiodrilus tamajusi* except that there happens to be a longer and fuller description of it than of many.

Fig. 1. *Notiodrilus tamajusi.* The worm shown from the ventral surface. About natural size (After Eisen.)

This earthworm is a native of Guatemala and is some six inches in length with a diameter of perhaps a quarter of an inch. The front part of the body is thicker than posteriorly. The body will be seen to be divided into some 218 rings by circular furrows which run right round the body. These divisions are

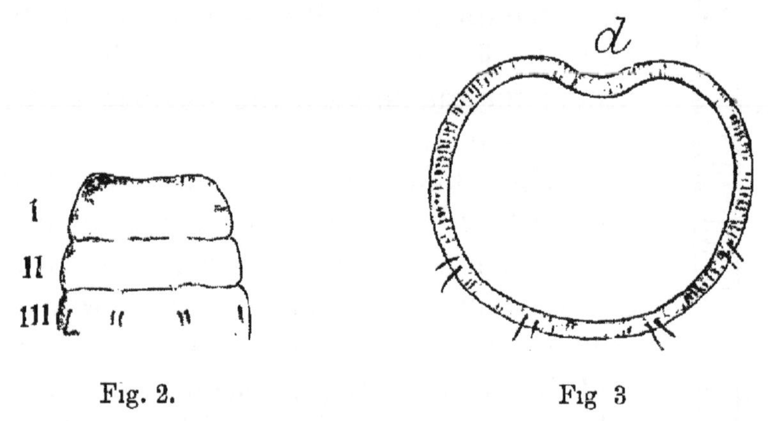

Fig. 2. Fig 3

Fig 2. The same species, first two segments and prostomium shown from ventral surface, I prostomium, II first segment without any setae, III second segment with its four pairs of setae (the dorsal-most seta of each outer pair is not visible in this view)

Fig 3. A section through the body of the same species showing the ventral position of the pairs of setae. (After Eisen)

termed segments or somites. At the head the mouth is surrounded by the first of these, and on the dorsal surface of that segment is a projection like an incomplete segment which is known as the prostomium. From the XIIIth segment to the middle of the XXth the body has a different appearance, and this region is known as the clitellum. Each of the 218 segments of the body except the first, and possibly one or two at the hinder end, is furnished with eight minute projecting bristles, the setae; these are disposed in pairs and all lie upon the ventral aspect of the worm. The movement of these by special muscles aids in locomotion.

An examination with even a hand lens shows a number of external pores which are important. Anteriorly there is the mouth which is overhung by the prostomium referred to above. At the extreme hind end—and surrounded by the last segment of the body—is the vent. Along the middle line of the back are a series of pores, one just at the very anterior edge of each segment, through which, when the worm is dried and then slightly pressed, liquid is seen to be ejected. These are called the dorsal pores and they belong one to each segment with the exception of the first seven, or—in some cases—more, segments. In front of one or other of the pair of setae which is situated most laterally, *i.e.* furthest from the ventral median line, is an orifice on each side in all but the first one or two segments of the body. These paired pores are the external outlets of the excretory organs frequently termed on account of their regular repetition with the segments 'segmental organs,' but more conveniently to be named nephridia. In the clitellar region and in fact on each of the segments XVII, XVIII, XIX are a pair of pores of which those on the XVIIIth segment are the least conspicuous. The large pair of pores on each of segments XVII and XIX occupy the position of the ventralmost pair of setae, which are here absent, or rather replaced by a very long curved and ornamented seta, which projects out of the orifice. These two pairs of pores are the outlets of the prostatic glands as they have been termed. The

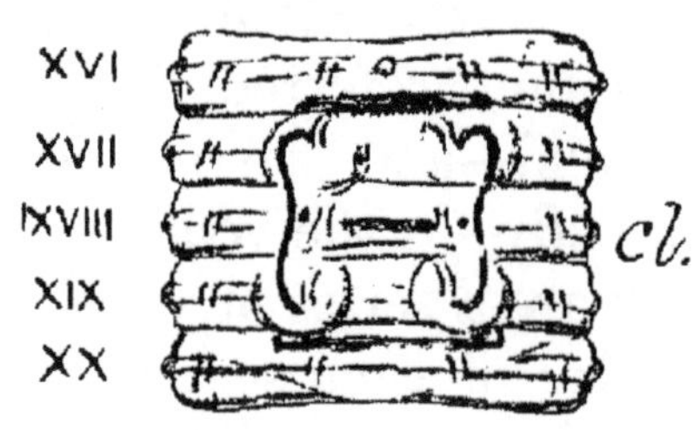

Fig. 4. The same species. Ventral view of segments XVI–XX (numbered in the figure) which form the clitellum, the posterior boundary of which is shown by a curved line on segment XX. The figure will be understood from the annexed description.

minute pair of pores on segment XVIII do not take up the position of the ventral setae; for these are present and to the inside of each pore. A groove, shaped something like a reversed 3 or the Greek letter Σ, connects the orifices of each side of the body, the middle part of the groove, where the two semicircular halves of which it is composed meet, coinciding with the minute pores on segment XVIII which are the orifices of the sperm ducts.

On segment XIV are a pair of very minute pores a little in front of the ventralmost setae and thus very near together. These are the openings of the oviducts. Finally, near to the anterior border-line of segments VIII and IX and on a line with the ventral pair of setae is a pore on each side through which the cavity of the spermathecae reaches the exterior.

So much then for the external characters of our worm. We next turn to the internal anatomy. When the worm is opened by a longitudinal section from end to end, and the two flaps of skin are turned outwards and pinned down, the internal structure is almost completely revealed. Running from end to end is seen the alimentary canal; the general cavity of the body (coelom) in which it lies, as do of course the other organs to be enumerated, is seen to be divided by cross divisions, the intersegmental septa, into a series of chambers which correspond with the external division into segments. The septa are in fact inserted on to the body-wall along the furrows which mark the divisions between adjacent segments. Anteriorly the large pharynx is responsible for confusing the arrangement of the septa, which become subdivided and fused or are prolonged a greater way backwards and thus present a less obviously segmental disposition. Certain of the more anteriorly placed of these septa are much thicker than the rest. This is the case with the septa which separate segments V to XII. The alimentary canal is perfectly straight and runs in the middle line, being supported by the septa which it perforates. The mouth leads into a buccal cavity which later becomes the pharynx, a portion of the tube which is much thickened by muscular walls dorsally. Then follows a very short section of the oesophagus and in the fifth segment this becomes the gizzard, a very characteristic organ with thick muscular walls quite smooth and with a very thick lining of structureless membrane. After this is a narrower tube, the rest of the oesophagus. Into this open in each of segments VII, VIII, IX a pair of calciferous glands; these are diverticula of the gut with much folded walls, the cells of which secrete carbonate of lime. In the XIIth segment or so, the oesophagus suddenly widens out to form the intestine which runs as such to the end of the body. This wider tube has a ridge running along its dorsal side, the typhlosole. Along the dorsal surface of the intestine and the oesophagus is seen a red tube, contractile during the life of the worm, which is the dorsal blood vessel and whose contained blood is coloured red, as is the blood of vertebrated animals, by haemoglobin. But in the earthworm the colouring matter is not situated in corpuscles as in the vertebrate. The dorsal vessel is connected by a few pairs of equally contractile transverse trunks with a ventral vessel which is not

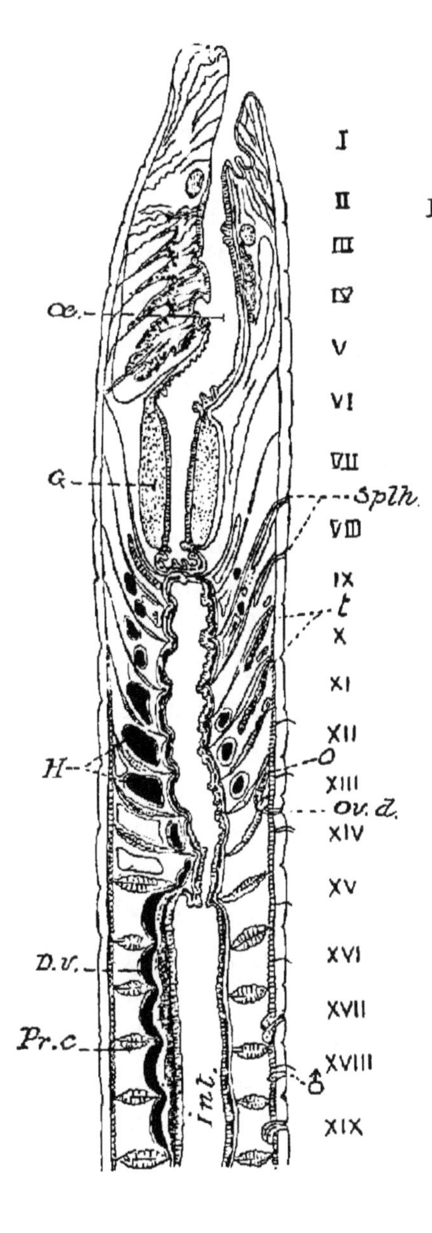

Fig. 5. A longitudinal section through the middle of the first nineteen segments of the body of *Notiodrilus vasliti* (a species very closely allied to that described in the text), the segments are numbered I, II, &c. *D.v.* dorsal blood vessel, *G* gizzard, *H* hearts, *œ* oesophagus lying in front of gizzard behind which another tract of still narrow oesophagus is seen opening into *Int.* intestine. The whole alimentary canal is supported by the intersegmental septa (*Pr.c.*) between which is the system of spaces forming the coelom, *ov.d.* pore of oviduct, *Splh.* orifices of spermathecae, ♂ orifice of sperm duct, *o* ovary, *t* spermaries. (After Eisen.)

contractile. There are other branches of these main longitudinal trunks and some minor longitudinal trunks which we shall not stop to describe further. The nervous system of the worm consists of a pair of ganglia which lie above the gut in the third segment; they are connected by a commissure running round the gut with a chain of closely fused pairs of ganglia, one for each segment to the very end of the body. In each of the segments, except the first two or three, there are a pair of excretory organs known as nephridia; these are essentially coiled glandular tubes opening on to the exterior by the regularly placed pores already referred to in considering the external characters. The tube ends in a funnel-shaped, and therefore dilated, mouth, which opens into the segment in front of that which contains the rest of the organ; a nephridium therefore lies in two segments. The only other important organs which are left for consideration are those devoted to the reproduction of the species. The essential organs are the spermaries and the ovaries. Of the former there are two pairs of minute whitish bodies which lie in segments X and XI on either side of the nerve cord attached to the anterior septal wall of their segments. The ovaries are not in the following, but in the XIIIth, segment, and occupy an identical position in that segment. A short tube with a funnel or trumpet-shaped and wide orifice opens into the cavity of the

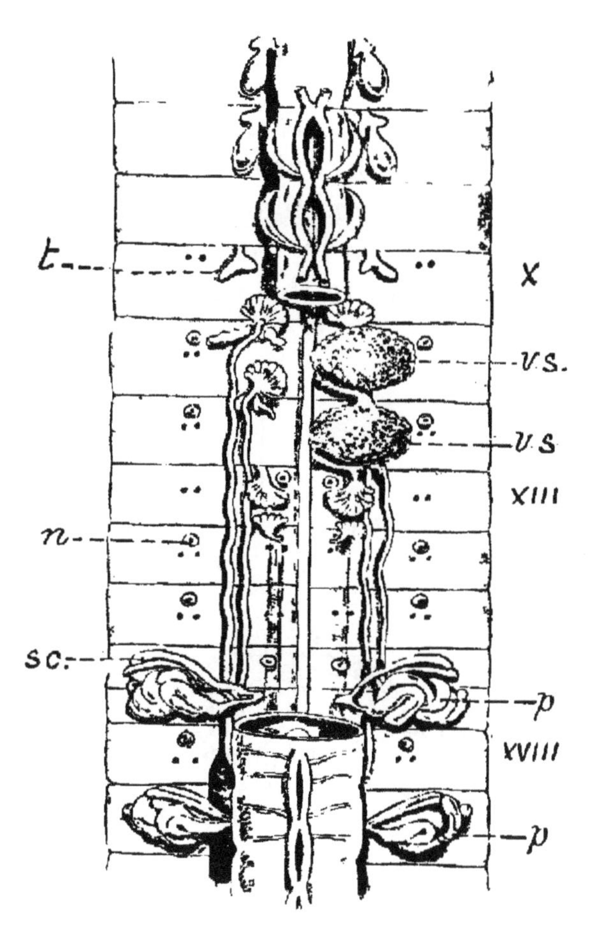

Fig 6. Genital region of *Maoridrilus dissimilis* (in which species the different organs are essentially as in *Notiodrilus tamajusi*). The oesophagus is cut away between the xth and xviiith segments to display the various organs. *n* external orifices of nephridia which alternate in different segments; the one lettered opens in front of dorsalmost pair of setae, that of segment xvi opens in front of ventralmost pair, *p* coiled glands opening on to xvith and xixth segments, *sc* sacs containing long seta, associated with these glands, *t* spermary of segment; another pair in an exactly similar position in segment xi. Behind spermaries are funnel-shaped openings of sperm ducts which are seen running along the body to their external orifice on xviiith segment. *v s* sperm sacs behind the posterior of these and in segment xiii is seen the large ovary corresponding in position to the spermary and opposite to each ovary the oviduct (× 3.)

XIIIth segment opposite to each, and, perforating the septum, opens on to the exterior on the XIVth segment. A similar but larger and more folded pair of trumpet-shaped funnels opens in the same way opposite to each spermary. But in this case the two tubes of the

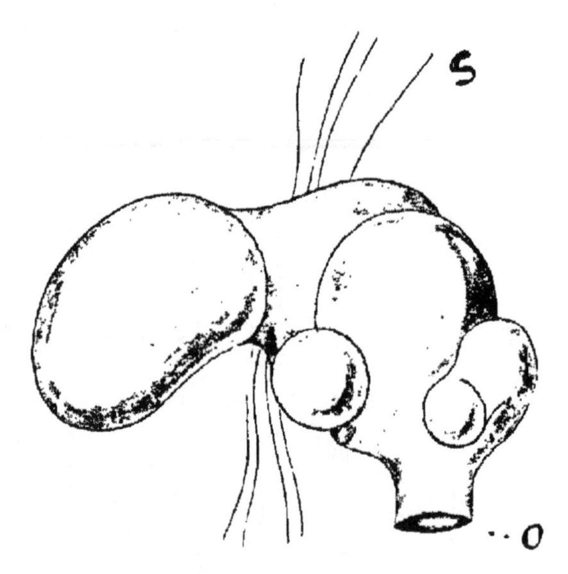

Fig. 7. A spermatheca of *Notiodrilus tamajusi*. The external orifice is shown at *o*; above this are the diverticula, *s* is a portion of an intersegmental septum. (After Eisen.)

sperm ducts run backwards for some way and those of each side after joining open on to the XVIIIth segment by the pores already mentioned. On the XVIIth and XIXth segments open two glands which are long and tubular in form and much coiled. These are the spermiducal glands and each opens in common with a muscular sac containing the long and ornamented seta referred to in describing the various external orifices. It will be noticed that the sperm duct has no direct connection with these glands but only indirectly through the external gutter which connects the three male orifices of each side of the body. Segments IX–XII inclusive contain certain sacs which depend from, and are formed as outgrowths of, the septa of those segments. These are the sperm sacs in which the male germ cells undergo their development. A corresponding body (but very much smaller) is sometimes found in relation to the ovary but has not been actually described in the particular species dealt with here. Finally, in segments VIII and IX are a pair (that is four altogether) of roundish sacs, with two or three minute diverticula, known as the spermathecae. In the diverticula of these sacs are stored the sperm derived from another individual.

This completes the general sketch of the structure of *Notiodrilus tamajusi* which we have selected as a type. In this same genus are a large number of species which differ from that selected in various small structural points. Thus in *N. annectens* (Beddard), a species from New Zealand, the spermaries and ovaries are attached to the posterior, instead of to the anterior, wall of their segments, and there are neither calciferous glands nor modified setae upon segments XVII and XIX.

In all essentials however the two types agree and are thus to be looked upon as referable to the same genus. Starting from the structure of these types we may now sketch in quite a brief way the main divergencies of structure shown in the group of Oligochaeta.

We shall naturally begin with the family Megascolecidae of which a type has just been described.

Within the limits of the same sub-family as that which contains *Notiodrilus*, *i.e.* the Acanthodrilinae, the changes of structure affect all the principal organs of the body except the nervous system, but are not very large and vary from genus to genus. They are mainly perhaps in the direction of reduction and simplification. Thus in *Chilota*, *Maheina* and *Yagansia* the spermaries are reduced to one pair in either the Xth or XIth segment, while in *Yagansia* one pair of spermathecae and of spermiducal glands have also disappeared. In *Microscolex* the spermaries remain normal, but one pair of spermathecae and of spermiducal glands have disappeared, the remaining organs of these series being in the IXth and XVIIth segments respectively. In *Microscolex*, *Chilota* and *Yagansia*, moreover, there is a further degeneration in the disappearance of the calciferous glands. These glands are often absent and sometimes less developed in the New Zealand *Maoridrilus*, which is otherwise not a degenerate form and differs characteristically from *Notiodrilus* by the fact that the paired nephridia alternate in position in successive segments, being now in front of the dorsal, and in other segments in front of the ventral, pairs of setae. While these genera are somewhat degenerate, the New Zealand *Plagiochaeta* has undergone specialisation in an upward direction. For the setae of each segment are increased to a large number much exceeding eight.

It is not a long step to the sub-families Diplocardiinae and Trigastrinae. In the first of these, an American race confined to the northern and central parts of that continent, the male pore shows a tendency to move backwards, being situated on any of segments XVIII–XXI. The two spermiducal glands follow it, but are always placed one pair in front and one behind, as in *Notiodrilus*. In this group we get a new feature of specialisation in the duplication or triplication of the gizzard.

So too with the Trigastrinae where there are either two or three gizzards; but in this sub-family another modification has become apparent. The paired nephridia have disappeared and their place is taken by several, often quite numerous, pairs of much smaller nephridia called on that account 'micronephridia' instead of 'meganephridia.' To this sub-family belong the especially African but also American and Malayan *Dichogaster*, whose name is derived from the important fact that it possesses two gizzards.

Not far off is to be placed another sub-family, that of the Octochaetinae, which is New Zealand and Indian in range, the intermediate countries being, strange to say, not populated by this race of Oligochaeta. The group contains several genera of which *Octochaetus*, *Eutyphoeus*, and *Dinodrilus* are the best known. All these worms agree in the main features of their anatomy with *Notiodrilus* ; but they have diverged in different directions. Thus in *Octochaetus* the typical two pairs of gonads and glands belonging to the generative system have been retained, while the nephridial system consists of micronephridia ; in *Eutyphoeus*, one pair of spermiducal glands has disappeared, and as a general rule the species of this genus have only one pair of spermaries and the corresponding pair of sperm ducts. They are close to *Octochaetus*. The third genus mentioned, *Dinodrilus*, is a New Zealand form specialised in possessing 12 setae in each segment. Otherwise it is not far removed from *Octochaetus*.

A fifth sub-family is also easily referable to the type whose structure has been dealt with as a preliminary to the present survey. That sub-family is the Ocnerodrilinae which is American and African in range. These worms are somewhat degenerate in comparison with their allies. Thus the calciferous

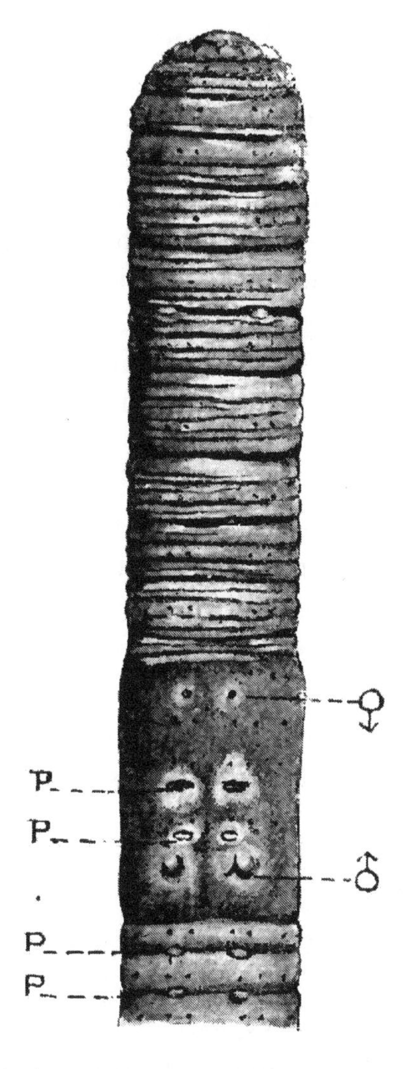

Fig. 8. Ventral view of *Eutyphoeus masoni*. *p* papillae, ♂ male pores, ♀ oviduct pores. (×3.)

glands are reduced to a single pair or to a single gland in the IXth segment, the nephridia though regular and paired have no covering plexus of blood vessels, and the worms themselves are slender and delicate, being indeed often aquatic in habit. The spermiducal glands, which are as in the former sub-families independent of the sperm ducts though sometimes opening in common with them into a short pocket-like ingrowth of the skin, are reduced in their minute structure and much simpler than in the other types.

The genus *Kerria* is the least reduced perhaps. It has the male pores on segment XVIII with a pair of spermiducal glands on the segments preceding and following this in the typical Acanthodriline fashion. There are two pairs of spermathecae in VIII and IX, but the spermaries are reduced to one pair in X. The gizzard is present. *Ocnerodrilus* is a little further reduced from this last. The gizzard has gone; there is but one pair of spermiducal glands (as a rule) opening in common with the extremity of the male duct on to segment XVII; the spermathecae also are reduced to one pair, but there are two pairs of spermaries. The African *Nannodrilus* is more robustly built. There are two gizzards, the male duct opens into a muscular pouch, into which also open one of the two or three pairs of spermiducal glands. There are two or three other genera and sub-genera not showing any great divergencies from the range of structure indicated in briefly defining those enumerated above.

Finally, we have the largest of all the sub-families of this family, viz. that of the Megascolecinae. These worms are mainly tropical in range and also mainly found in the Indo-Australian portion of the world. In them the sperm ducts open in common with the usually single pair of spermiducal glands and prevalently upon the XVIIIth segment. The glands moreover have not always, though they often have, the tubular form shown in all the sub-families hitherto considered. In many forms they are branched and lobate glands, and if there are two pairs one may be of one type and the other of the second and derived type, as for instance in *Megascolex ceylonicus*. Furthermore, it is much commoner among the genera of this sub-family for the setae to become numerous and to spread right round the segment; this condition is seen in the genera *Pheretima, Megascolex, Diporochaeta, Perionyx, Plionogaster*. The spermathecae also are commonly more than the typical two pairs of the forms already considered, and in certain species (for example *Pheretima hexatheca*) there are as many as six pairs of those organs which are moreover—and in this they resemble the majority of species of the last sub-families—nearly always furnished with a diverticulum or diverticula. The nephridia are either paired or numerous and these various characters allow of the sub-family being split up into sixteen genera or thereabouts.

As an example of another type of organisation and as contrasting with *Notiodrilus* we may now briefly describe the structure of the genus *Pontoscolex* (better known as *Urochaeta*), a member of the American and African family Geoscolecidae.

The worm is some four inches long and composed of as many as 212 segments. Each of these except the first has eight setae which for the first few segments of their occurrence are disposed in four pairs in the usual fashion. Behind this point however the setae cease to present this symmetrical arrangement and are irregularly disposed so that a given seta is not in the same line with the corresponding seta of the segments in front or behind. There is thus no region of the body which has not a seta implanted upon it; and the effect is therefore comparable to the condition obtaining in those worms, such as *Pheretima*, where circles of numerous setae are met with. There are however only eight in a given segment. The clitellum extends from segment XV to XXII or XXIII and is developed only dorsally and laterally. It has setae like the rest of the body; but those upon the clitellum are longer and more markedly ornamented than are those of the body generally. The latter are not only sculptured with fine ridges but are bifid at their free extremity. The prostomium is often apparently completely absent. It is however really present but is retractile. As to the pores which are visible externally the dorsal pores are completely absent. The pores of the nephridia lie in front of the dorsal pair of setae or in a line corresponding to the position of those setae where the arrangement has become irregular. The spermathecal pores are three pairs and are in the very front of segments VII, VIII, IX. The male pores, very inconspicuous, lie on the ventral side of segment XXI just within the clitellum. The oviducal pores are on segment XIV.

As to internal anatomy the general plan of the segmentation shows no great differences. Certain septa only show a difference, *i.e.* those separating segments VI–XI which are specially thickened. In the alimentary canal a gizzard in segment VI is to be noted and three pairs of calciferous glands in segments VII–IX. The nephridia are paired structures and commence early. The first two or three segments are occupied by a pair of large glands opening into the buccal cavity which are apparently a slightly modified pair of nephridia and are generally termed 'peptonephridia' since they are associated, as it would appear, with the function of alimentation and are not purely excretory organs. There is but a single pair of spermaries in segment XI, and of ovaries in segment XIII. The sperm ducts open on to the exterior in the position already mentioned and they are not associated at their pore with any glands comparable to spermiducal glands. A pair of sperm sacs depend from segment XI and traverse a considerable number of segments, being thus long and tongue-shaped instead of short and limited to one segment. The spermathecae are three pairs of elongated sacs in segments VII–IX, without any diverticula at all.

It will appear therefore that many and considerable differences divide *Pontoscolex* from *Notiodrilus* and indeed from all of the Megascolecidae whose structure has been touched upon in the foregoing pages. The most important of these are the ornamented setae and their arrangement and the modification of the setae upon the clitellum: the absence of diverticula to the spermathecae: the absence of terminal glands associated with the male ducts. Although taken in their entirety these characters are distinctive of the American Geoscolecidae (sub-family Geoscolecinae), there is no one of them which is not to be found in some Megascolecid. Thus the subgenus *Ilyogenia* (of *Ocnerodrilus*) has sometimes no spermiducal glands: the genus *Perionyx* has spermathecae without diverticula in some species, and other genera of Megascolecinae are in a like condition. The setae of *Dichogaster* are sometimes ornamented, while in *Pheretima houlleti* the clitellar setae are different from those upon the other segments.

We can in fact only define the family Geoscolecidae by an assemblage of characters which are mainly these: dorsal pores absent, only a few in the neck region being occasionally present; setae generally ornamented, those of the clitellum being larger and more marked than the others; spermathecae without diverticula; often instead of a pair of those pouches in the segment a large number of very small sacs, as in *Microchaeta, Kynotus*. Sperm ducts without terminal glandular or muscular sac, except in a few cases; setae always eight in a segment except in the genus *Periscolex* which has acquired the 'perichaetous' condition. The range of variation shown in the anatomy of the Geoscolecidae will be best taken in connection with the several sub-families into which it has been subdivided. In the first of these, the Geoscolecinae, no great differences divide the genera from that selected as the type, viz. *Pontoscolex*. The long sperm sacs attain to an extraordinary length in *Trichochaeta* (or *Hesperoscolex*) where the single pair extends through no less than 109 segments. Though as a general rule the sperm ducts open directly on to the exterior they do so through the intermediary of a large pouch in *Glossoscolex* (= *Titanus*). In *Onychochaeta* the setae on the last segments of the body are very much enlarged and thus form a more effective means of holding on to the soil than is possessed by other species.

The sub-family Hormogastrinae which contains but a single genus *Hormogaster* is remarkable for being limited in range to the Mediterranean coasts. The genus is mainly distinguished by possessing three gizzards; otherwise it is not very different from the sub-family just described. The African and Madagascar forms are associated (together with a few Asiatic forms) into a third sub-family Microchaetinae. These worms frequently possess a considerable number of very small spermathecae in segments XII, XIII or thereabouts instead of the usual paired arrangement. They have too very often glands connected with the enlarged setae already mentioned which are however (in the genus *Kynotus* at any rate) usually in front of the clitellum. The latter organ, contrary to what we find among the Geoscolecinae, is often behind the point of orifice of the male pores. This is so with *Kynotus*.

The last sub-family, Criodrilinae, has but three genera *Criodrilus*, *Sparganophilus* and *Alma*. These worms do not show any very marked differences from other Geoscolecids. *Alma* is noteworthy for the facts that the male pores are borne upon long processes of the body which bear specially modified setae and that one species at any rate has gills.

Another type of structure is offered by the Eudrilid earthworms which form rather a restricted family. These worms are as a rule quite easy to

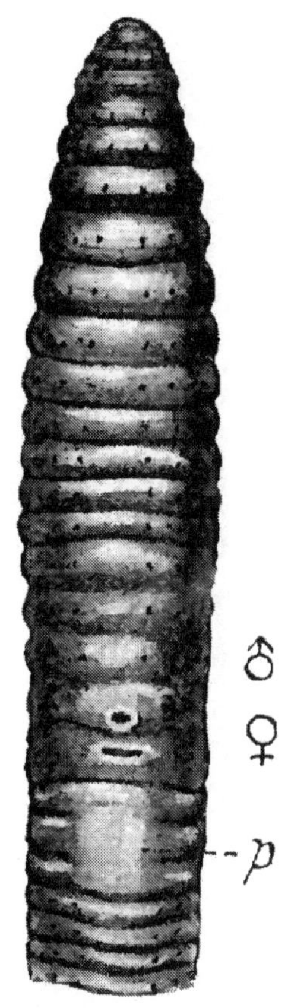

Fig. 9. Ventral view of *Polytoreutus Kilindinensis*, head end (×2). *p* papillae, ♂ male pore, ♀ female pore.

distinguish by their external characters only. For the apertures of the spermathecae and sperm ducts are apt to be very large and conspicuous. They are also generally unpaired, a character which is however not confined to the Eudrilidae; for there are Mega-

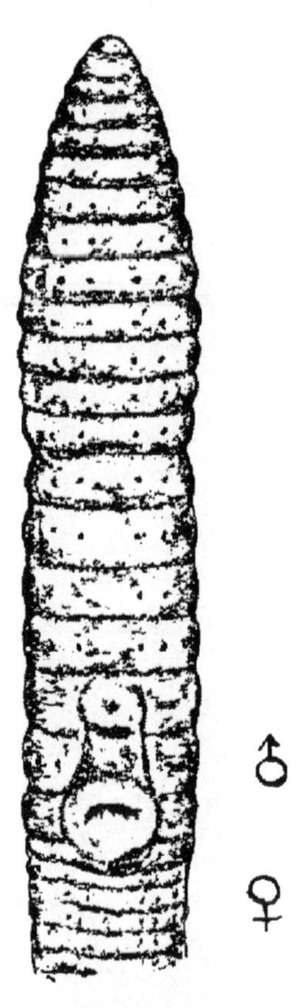

Fig. 10 Ventral view of *Polytoreutus finni*, head end (× 2), lettering as in fig. 9.

scolecids, such as *Fletcherodrilus*, and Geoscolecids in which the same unpaired character occurs. The principal feature of the family is that the ovaries are commonly enclosed in sacs—comparable to the sperm sacs which frequently envelop the spermaries in other

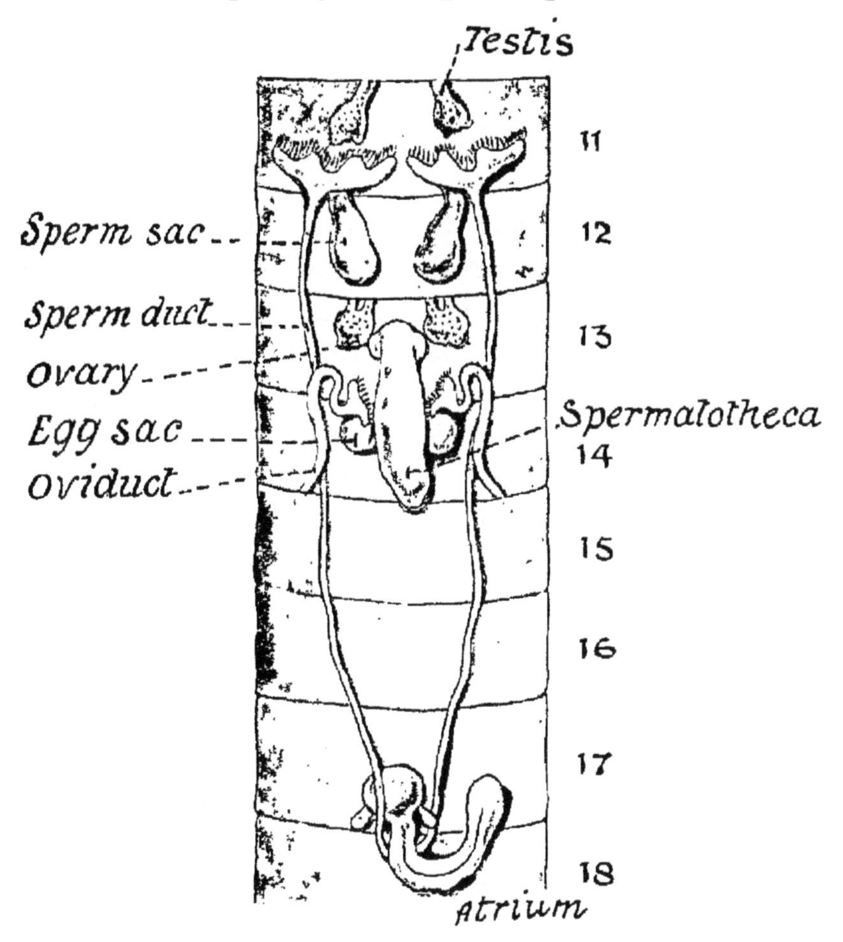

Fig. 11. Organs of reproduction of *Eudriloides durbanensis*.

earthworms—and that these sacs not only contain the mouths of the oviducts but are directly continuous with the single or double spermatheca. This is usually a large sac, always single or consisting of one pair only, which opens on to the exterior close to the oviducal pores; these spermathecae in the Eudrilidae are not comparable to the spermathecae of other earthworms; for they are in a way comparable to the sperm sacs, being formed as outgrowths of the septa. There is some variation of structure within the family. In a number which are associated into a sub-family Eudrilacea there are two paired calciferous glands and a single unpaired one, while the paired nephridia open by a large pore on to the exterior. In a parallel sub-family, the Pareudrilacea, the calciferous glands are apt to be more numerous and have a totally different structure: they have been apparently converted into non-digestive glands bearing some relation to the vascular system. The nephridia moreover do not open on to the exterior by single pores, but form a network within the thickness of the body wall and then open by numerous pores. There is however no resemblance here to the micronephridia of *Dichogaster* and other Megascolecids. In *Libyodrilus* (as an example of the Pareudrilacea) each nephridium forms a network out of the duct leading to the exterior. In the interior of the body a series of paired meganephridia are visible.

The earthworms of Europe belonging to the family Lumbricidae offer again a rather different type of structure, which is more reminiscent of the Geoscolecidae than of the Megascolecidae or Eudrilidae. In this family there are no glands appended to, or in the neighbourhood of, the orifices of the sperm ducts, such as are found in the other forms. As in the Geoscolecidae the clitellum is furnished with setae somewhat different in form from those which deck the body generally. These setae are never more than eight in a segment. Dorsal pores (absent in Geoscolecidae and in Eudrilidae) are invariably present. The spermathecae are without appendices and nearly always simply paired, though rarely we get numerous much smaller spermathecae in a single segment, as in *Kynotus* among the Microchaetine Geoscolecids. Internally the most striking feature of this family is to be seen in the position of the gizzard at the end of the oesophagus and at the beginning of intestine. The apertures of the male pores are—save for two or three exceptions where they are further forward—invariably upon the fifteenth segment, and the clitellum, often very long, usually begins behind this point, features which are also seen in *Kynotus*.

Finally we have the Moniligastridae which differ from all the types hitherto considered in a few rather important particulars. These worms are named on account of the fact that they possess several gizzards upon the oesophagus, a character which is however met with in the Megascolecid genus *Plionogaster* and in certain Eudrilids, *e.g. Hyperiodrilus*. The main peculiarity of the family is that the sperm ducts are very short and open on to the next segment to that which contains the spermaries, as in the water-living Oligochaeta generally. The terminal sac into which the male ducts open is moreover rather like that of such a family as the Lumbriculidae.

THE AQUATIC FAMILIES OF OLIGOCHAETA.

It would seem to be quite possible that when the fresh waters of the world have been as well examined for Oligochaeta as have so many parts of the land areas, the number of purely aquatic Oligochaeta will be found to equal those inhabiting the land. In any case we are quite justified at the present moment in stating that there are rather more families of these smaller Oligochaeta than there are of the bulkier terrestrial forms. But while there are certainly seven or eight distinct families, these do not between them contain at present so many genera as do the fewer families of earthworms; and the number of species of the latter that are known to science enormously exceeds that of the 'Limicolae' as the fresh-water worms were at one time called in common. The fact that there are purely marine forms of these water worms was hardly appreciated at the time that the term Limicolae was used; now however we are acquainted with a few such forms, and even with some which live at will in either fresh, salt, or brackish water. Of these something will be said later.

These forms have also been collectively treated of as Microdrili, a term which expresses the undoubted fact that they are all of small size and sometimes even minute; others however reach the dimensions of the smaller species of earthworms. There are a certain number of characters shared by the various families which may be considered first of all, before dividing them into their several subdivisions. These aquatic Oligochaetes are usually tender and transparent, the muscular layers of the body wall being much reduced as compared with the tougher terrestrial forms. The clitellum is also thinner and consists of a single layer of cells only, thus contrasting with the double layered clitellum of earthworms. As a rule the alimentary tract is simplified, there being no gizzard or glandular appendices of the oesophagus comparable to the calciferous glands of most earthworms. But this rule is not without exceptions; for we find in *Haplotaxis* a gizzard occasionally developed, and in the remarkable genus *Agriodrilus* from the Baikal lake a continuous gizzard formation along the oesophagus, while the Enchytraeidae may show something very like calciferous glands: and even a Tubificid, called by Pierantoni *Limnodriloides*, has a pair of diverticula of the gut.

Other internal organs show certain points of likeness in all or in the great majority of the aquatic families. Thus the nephridia are without a plexus of blood capillaries surrounding them, a state of affairs which also occurs in some of the slender Ocnerodrilinae among the earthworms. These paired organs also are very frequently not found in the anterior segments of the body and these include also as a general rule the segments in which the reproductive elements are formed. Save for an exceptional case among the genera of Enchytraeidae the dorsal pores are not found among the Limicolae, but in some cases at least a single pore, the head pore, is found. The sperm ducts, which among earthworms usually (and indeed always save in the anomalous Moniligastridae) traverse a considerable number of segments on their way from the internal opening into the body-cavity to the external pore, do not show the same phenomenon among the Limicolous Oligochaeta. They are sometimes indeed limited to a single segment, that is to say the funnels and the external pore lie in one segment. In other cases they open on to the exterior in the segment next to that which bears the funnel, and it is only rarely that they traverse more than one segment. There are also points of difference of general applicability to be noted in the sperm sacs and egg sacs. The latter are large and extensive, which is not the case among earthworms, and the former are as a rule more extensive in the number of segments that they occupy than among the terrestrial forms. Another difference which they show is that their cavity is quite simple and not divided up by trabeculae into numerous intercommunicating chambers as in the earthworms. Finally the eggs of the aquatic Oligochaeta are large and full of yolk and thus contrast with the very small ova of the earthworms which are moreover much more abundant. These features are either of general or universal occurrence and together form an assemblage of characters which mark out the aquatic families of Oligochaeta from their, at least mainly, terrestrial allies.

We may also refer to certain structures which although not universal among these aquatic families are nevertheless found only in them—that is, are not found in any family of the terrestrial worms of this order. The most salient of such characters are the long and hair-shaped setae tapering to a fine point and often provided with a series of delicate branches like a feather; such setae are often of very great length and they occur in their various modifications among the Aeolosomatidae, Naididae, and Tubificidae. It is clear that these delicate setae, though they may not be due in any way to the aquatic life, are rendered possible by it. To drag such tender processes through stiff clay would surely break and tear them out. It may be also mentioned that among the aquatic families as a rule the intersegmental septa do not show that thickening in some of the anterior segments of the body which is so general a feature of the land-dwelling species. Finally it is only among the aquatic forms, and among them only in the families Aeolosomatidae and Naididae, that asexual reproduction by budding takes place. Indeed so common and usual is it in the genera of these families that even yet there are considerable lacunae in our knowledge of the organs of reproduction in the said families.

Together with these general similarities are many points of structural difference among the worms inhabiting ponds, lakes, and rivers, which allow of their being divided into a number of quite distinct families.

One of the most distinct of these families and lying in any case quite at the base of the series is the family Aeolosomatidae which includes a number of distinct species of delicate and transparent worms, and in whose integument are embedded round cells bearing a large brightly coloured oil drop; this is reddish or green in colour, or—and this more rarely—colourless, but still recognisable as an oil drop. The green sometimes even verges upon blue on the one side and yellow on the other, while the red may approach brown or purple. These worms are assigned for the most part to the genus *Aeolosoma* which is found in all of the great continents and of which seven or eight species are known. To a more doubtful genus *Pleurophleps* are assigned a few small worms which have the general appearance of *Aeolosoma*, but are without the coloured or colourless oil drops in the skin. These worms have a very large prostomium which is ciliated on the lower surface, and the body is not markedly segmented externally by constrictions or internally by septa. The bristles are slender and hair-like, but among them are in some species the shorter stouter bristles bifid at the free tip, which are so general among the aquatic families of the Oligochaeta. These worms are not uncommon objects in pools containing weed; and they are to be found usually crawling among the weed. They consist as a rule of but few segments to most of which a pair of nephridia belong. The ovaries and the spermaries are only known in a few forms and appear to be unpaired and lie respectively in the fifth and sixth segments. There are 1–3 pairs of spermathecae, and the sperm ducts if distinct from, are at least very like, nephridia. The ova appear to make their way to the exterior by a large aperture in the ventral middle line of a middle segment of the body. The vascular system contains uncoloured blood and is greatly simplified.

The next family to the Aeolosomatidae in zoological position is clearly the Naididae. These are also small worms, but show in some respects a higher grade of organisation than their allies. While asexual generation is general, the reproductive organs are more commonly found than in *Aeolosoma*, though there are still many hiatus in our knowledge of the same in certain genera. Where they are known it has been found that the spermaries and ovaries are very far forward in the body, in the fifth and sixth segments respectively. The spermathecae are in segment six and the male ducts open into a terminal chamber, called 'atrium,' which is on the whole not unlike that of the Tubificidae. The blood in these worms is red as in the higher types, and thus differs from that of the genus *Aeolosoma*. The setae are rather varied, being in some cases long and slender, sometimes greatly exaggerated in length as in *Ripistes*; other setae are forked at the free end, and in *Paranais* this is the only kind of setae met with. A marked feature of this family is that the dorsal bundles of setae do not always begin like the ventral setae upon the second segment of the body. Indeed in *Schmardaella* there are no bundles of dorsal setae at all. The Indian genus *Branchiodrilus* is remarkable for the fact that it has paired processes of the body which may be termed gills and which in some segments involve the dorsal setae. Another kind of gill is found in the genus *Dero* (which has many species) and in the allied *Aulophorus*. These are placed round the vent, and are ciliated. Other genera are *Nais*, *Chaetogaster*, *Vejdovskyella*, *Amphichaeta*, *Stylaria*, *Macrochaetina*, *Pristina*, *Naidium*.

Several genera, *Pristina*, *Nais*, *Dero*, are found in many parts of the world; but it is not possible at present to consider very seriously the facts of their geographical distribution.

Next to the Naids a group of aquatic worms present themselves for consideration which are usually placed in three distinct families, which families are however rather hard to define. These three families are the *Tubificidae*, *Phreodrilidae*, and *Lumbriculidae*. The Phreodrilidae were at one time placed with the Tubificidae by Michaelsen. It is now perhaps the general opinion that they form a family of their own, at any rate since the discovery of two other genera *Phreodriloides* and *Astacopsidrilus*, besides the original genus founded by myself, and named *Phreodrilus* from the fact that the species was found in a deep well in New Zealand.

The distribution of this family especially of the genus *Phreodrilus* is extremely interesting. The genus *Phreodrilus* is the only one genus of the aquatic Oligochaeta which has, like *Notiodrilus*, a circumpolar range, the pole being the south pole. It occurs in New Zealand, in Kerguelen, and, if we are to accept the opinions of Drs Michaelsen and

Benham that my genus *Hesperodrilus* is to be merged in *Phreodrilus*, in Patagonia also.

In this genus the male pores are upon the XIIth segment while the spermaries lie in segment XI. Moreover the spermathecae lie behind the male pores. In the Tubificidae on the other hand it is at least the rule for the spermaries and male pores to be pushed a segment further forwards. And the spermathecae are before the male pores. *Phreodriloides* is like *Phreodrilus* but has no spermathecae. It is also New Zealand in range. *Astacopsidrilus* is Australian and is semi-parasitic upon the Crayfish *Astacopsis*, whence its generic name. *Phreodrilus branchiatus* is one of the few forms of Oligochaeta that possesses gills. Of these there are a series of pairs on about the last eleven segments of the body. They are lateral in position, and thus contrast with the also gilled *Branchiura sowerbii*, where the gills, also on the posterior segments of the body, are more numerous and lie dorsally and ventrally, a pair to each segment.

The Tubificidae differ from the Phreodrilidae mainly in the points already noted. There are a considerable number of genera of which the following are the best known, viz., *Tubifex*, *Limnodrilus*, *Limnodriloides*, *Branchiura*, *Lophochaeta*, *Ilyodrilus*, *Psammoryctes*, *Clitellio*, *Telmatodrilus*, *Bothrioneuron*, *Lycodrilus*.

The Tubificidae are mainly northern temperate forms, and a few of them such as *Clitellio arenarius* and '*Peloryctes inquilina*[1]' are found on the sea coast. There are also a few of this family in the southern hemisphere. These forms include *Clitellio abjornseni* from Australia, and a few species of *Branchiura* from New Zealand and the islands of the Antarctic ocean. There is also to be mentioned *Rhizodrilus* (or *Vermiculus*) *aucklandicus* from the island of that name in the New Zealand area. The only tropical species appears to be the Indian and Malayan *Bothrioneuron iris*, though this genus also occurs in Europe and in southern South America. It is quite likely however that *Branchiura sowerbii*, a species known at present from tanks in hot houses, may be a tropical American species.

The family Lumbriculidae is yet more restricted in its range. It has not yet been met with away from the temperate northern hemisphere, and the great variety of species recently described from Lake Baikal by Dr Michaelsen is a very remarkable fact. The Lumbriculidae are entirely fresh water in habit and not even partially terrestrial. The following are the principal known genera: *Lumbriculus, Trichodrilus, Rhynchelmis, Phreatothrix, Claparedilla, Stylodrilus, Mesoporodrilus, Sutroa, Eclipidrilus, Aurantina, Athecospermia, Lamprodrilus, Teleutoscolex.*

[1] With many synonyms, including *Tubifex ater* (see p. 53).

In the worms of this family the male pores are usually upon the tenth segment but sometimes upon the eighth or eleventh. And in addition to a pair of funnels in the antecedent segment there is also commonly a second pair in the same segment as that which contains the external pore. The two sperm ducts however open into the same distended atrial cavity before opening on to the exterior. In *Lamprodrilus* however each sperm duct opens by its own separate atrium on to the exterior and in two consecutive segments. In *Teleutoscolex* there is but one pair of funnels opening into the same segment with the atrial pore.

Near perhaps to the Lumbriculidae comes a very small family indeed, that of the Alluroididae. So small is it that it consists of but a single genus *Alluroides* of which there are two species *A. pordagei* and *A. tanganyikae.* Both of these species were described by myself, one of them from but a single specimen, the other from only two. Both species—and the name of one denotes the region—are from eastern tropical Africa. The Tanganyika worm is purely aquatic; the other species was found in a swamp of the mainland opposite to Mombasa. This remarkable genus has a pair of spermaries in segment X; but the ovaries are as in earthworms in the XIIIth segment. The male pores are upon that segment also, *i.e.* further back than in the aquatic worms. The spermathecae open close to the median dorsal line of the body in one species; in *A. tanganyikae* there is but one spermatheca which opens actually in the dorsal median line between segments VIII and IX. This family seems to come nearest to the Lumbriculidae but has also obvious points of likeness to the terrestrial Moniligastridae. It fully deserves a separate family, which was founded for it by Dr Michaelsen.

Not obviously related to any of the other families of Oligochaeta is the family Enchytraeidae. This consists of a very large number of species, which are placed in about a dozen genera, and whose habitat is nearly confined to the cold and temperate regions of the world. A large number of species for example have been described by Dr Eisen from Alaska, while others inhabit the verge of Patagonia. It is only a few which are found in warmer countries. There is for instance a solitary Enchytraeid in India and the neighbouring parts of Asia described by myself as *Henlea lefroyi* but doubtfully of that genus according to Dr Michaelsen. I have also myself obtained another Enchytraeid from British Guiana. Apart from such rare exceptions the family is arctic or temperate in its range and is even met with upon the ice of glaciers. These little worms—they are very rarely of more than minute size—are both aquatic and terrestrial and inhabitants of the sea shore. They have four bundles of short often straight and somewhat stumpy setae; *Anachaeta* is entirely without setae. That they bear some relation to the lowest group of Oligochaeta, that of the Aeolosomas and Naids, appears to be shown by the very anterior position of the spermathecae, *i.e.* in the fourth or fifth segment. The spermaries and ovaries on the other hand are in segments XI and XII. They are peculiar among the aquatic families in having complex glands appended to the oesophagus which recall the calciferous glands of the earthworms. The funnel of the sperm duct is extraordinarily deep and lined with glandular cells except in an apparently primitive genus from Lake Baikal.

The remaining family of the Limicolae is that of the Haplotaxidae which contains two genera, viz., *Haplotaxis*, better known as *Phreoryctes*, and *Pelodrilus*. These two genera overlap somewhat in their characters and it is in the present state of our knowledge rather difficult, if indeed possible, to differentiate them thoroughly. They are slender worms which frequent either damp earth or water and are thus somewhat intermediate in habit between the Limicolae and the Terricolae. The chief peculiarity of their structure lies in the fact that the sperm ducts are unprovided with any kind of terminal apparatus whatever, but open directly upon the exterior. Moreover there are generally two pairs of testes in segments X and XI, and in some species two pairs of ovaries in the two following segments. The small family is very widely distributed in more temperate regions, principally of the antarctic hemisphere. It occurs for instance in New Zealand, South Australia, the Cape, and in the northern hemisphere in Europe, Western Asia, and North America.

CHAPTER II

MODE OF LIFE

We have now completed a brief survey of the general characters of the group of the Oligochaeta which will at least serve to impress upon the reader the fact that these animals are somewhat diverse in structure, and that even as regards outward appearance it is not difficult to distinguish a large number of different types. These facts become all the more remarkable when we reflect upon the very similar conditions which surround all the species of earthworms. It is not clear how far the influence of the soil differs in a tropical forest in South America and in Central Africa. With divergent external conditions anatomical differentiation becomes more accountable. But in this case we have a lavish anatomical variation which would appear to have no connection with any kind of need that we can as yet appreciate. Comparing the Terricolous Oligochaeta with other large groups of the animal kingdom, all or nearly all the members of which lead a closely similar life, such as the frogs and toads, or the land mollusca, or snakes, we get a much wider range of structural change in the Oligochaeta than in any of these.

We shall now consider their mode of life and their relations to the environment.

The mode of life of earthworms seems at first sight to need no special chapter or section. They simply live in and beneath the soil, leaving it at times to wander over the surface especially at night and during wet weather. But there are a number of species which habitually lead an aquatic life and whose characteristics therefore demand consideration.

Aquatic Earthworms.

Although it is perhaps somewhat of a contradiction to speak of aquatic earthworms the collocation of words will serve to emphasise the fact that there are species of Oligochaeta belonging to the tribe Megadrili or terrestrial group, which are as purely aquatic in their habits as is a *Tubifex* or *Limnodrilus*. In such cases we may fairly assume rather a return to an aquatic life than the persistence of such a habit. For we do not find among these genera and species much evidence of particular resemblance with the purely aquatic familes of Oligochaeta. It is therefore particularly interesting to examine into the characteristics of these water-living genera; for we may expect to be able to deduce from them some hints as to what characters are really to be associated with the purely aquatic life. We can in fact hope to differentiate between adaptive and fundamental characters in these animals.

These secondarily aquatic species can be referred to two categories. There are examples of particular species which differ from their congeners in being aquatic; and there are whole genera, even sub-families, which are exclusively, or very nearly so, aquatic in habit. The former division need not detain us; for the actual occurrence of the worms in fresh water instead of upon dry land may be a temporary affair and not a mark of habitual sojourn. Thus I myself found the British and European earthworm *Eiseniella* (*Allurus*) *tetraedrus* in the River Plym in Devonshire, while it has been generally met with upon land. The Patagonian and Falkland Island species *Notiodrilus aquarum dulcium* was so called on account of its having been collected in fresh water. But its near ally *N. georgianus* (which is perhaps even identical with it) was found on the sea shore in the same region of the world. While the differences which the small species of *Notiodrilus* shows from other purely terrestrial members of the same genus are trifling, further information may prove that this case is on all fours with that of *Eiseniella* referred to above. There are plenty of similar instances which I shall not enumerate.

We may now therefore pass on to the second category. These examples are obviously much more important because they are of worms which appear to be wholly aquatic, or very nearly so, and which belong to definite genera easily distinguishable as such from their allies. The examples are not however very numerous. And they belong practically exclusively to the family Geoscolecidae, a family which, it will be seen later, is confined to South America, South Africa, Madagascar, certain parts of India and Burmah and of Europe. It is not a family which has reached the greater part of the East or which has been carried to the Antarctic parts of the globe. It is furthermore very important to bear in mind that there are reasons for regarding this family Geoscolecidae as one of the more modern branches of the Oligochaeta; this latter statement tends to prove that the aquatic life is, as already suggested, a secondary matter in these worms, and is not due to their belonging to an ancient race which has never left the waters of the land.

A very interesting fact offers itself first of all in considering this family of earthworms. The Geoscolecidae are one of the only division[1] of the Oligochaeta terricolae which are generally found to be without those characteristic series of pores in the middle line of the back known as the dorsal pores. They are thus eminently suited for an aquatic life; for it is to be supposed from the fact that the purely aquatic 'Limicolae' are also without these pores that their existence is prejudicial to a water-living worm. Indeed the entrance of water into the body-cavity would presumably be dangerous to the worm. The Geoscolecidae are thus already marked out, as it were, for an aquatic life. No modification is here necessary for them. It is also to be noted in this connection that various species of the genus *Notiodrilus* to which reference has been made as a partly aquatic genus have no dorsal pores. They too are thus fitted for at least an amphibious life.

The rule however regarding the absence of dorsal pores in the Geoscolecidae is not absolute. A few species and among them two species at any rate of the aquatic genus *Sparganophilus* have a few pores between some of the anterior segments which have been spoken of as 'neck pores.' They are not, it is to be believed, of a different nature from the generally distributed dorsal pores of other worms but are in fact limited to the 'neck' region.

There are no other obvious characters of the

[1] In the Eudrilidae also these pores are very frequently absent.

family Geoscolecidae as a whole which might be regarded as fitting them for a purely aquatic life.

Of this family one entire sub-family, the Criodrilinae, is very nearly purely aquatic in habit. Two genera, viz. *Callidrilus* and *Glyphidrilus*, out of another sub-family, Microchaetinae, which contains perhaps five other genera, are also aquatic in their mode of life. In examining into the characters of the various aquatic species with a view to searching for common characters which might be put down to modifications induced by the aquatic life, there are two or three which arrest attention. In the first place the Criodrilinae never possess a well-developed gizzard, having at most a rudimentary gizzard, or even two. However this character is not of universal applicability, for both *Callidrilus* and *Glyphidrilus* have got a gizzard and a strong one. These later genera however have no calciferous glands or oesophageal pouches of any description, structures which are also absent among the Criodrilinae. It will be remembered that the purely aquatic families, Tubificidae, Lumbriculidae, etc., rarely show signs of a gizzard and rarely possess oesophageal pouches of any kind. In view of the fact that in the case of a life in fresh water no calcareous salts are necessary to resist the acids of the soil, and that the mud passed through the alimentary canal is already finely divided, it is not surprising to find gizzard and calciferous glands absent or rudimentary.

Another not unusual feature among these aquatic Geoscolecidae is the quadrangular form of the posterior end of the body. This is shown—as its specific name denotes—by *Glyphidrilus quadrangulus*, by species of *Alma* and in all the species of the genus *Criodrilus*. It is to be noted in this connection that a species of the partly aquatic *Eiseniella* has been named *tetraedrus* on account of precisely the same phenomenon. In these cases it is the posterior part of the body which is thus quadrangular; the anterior segments down to the ninth in *Criodrilus* being rounded in the usual Oligochaetous fashion. As the paired setae are apt to lie in the four projections of the quadrangular body, one is tempted to see in this arrangement of structures a faint approach to the dorsal and ventral parapodia of the marine worms, and in any case it seems possible that by this means the worms can cling more effectively and continuously to the stems and leaves of aquatic plants among which they so largely live.

It is a very remarkable fact that in the genera *Criodrilus* and *Alma* the vent is dorsal in position instead of being surrounded as in earthworms generally by the last segment of the body. This fact might be put down to the near affinity between these two genera, were it not for the fact that *Glyphidrilus* shows precisely the same state of affairs. These facts gain additional significance in my opinion from the fact that among the leeches which are now universally admitted to be allies of the earthworms the same position of the vent is met with. This abnormal position of the end of the alimentary canal may thus be fairly quoted in connection with structures modified by the aquatic life.

Finally, and this seems to be very important, the only genus among the Megadrili which possesses gills is the Nile worm *Alma nilotica.*

Marine Species.

There are a few, but relatively speaking very few, worms of the order Oligochaeta which lead a marine life. And of these the majority are shore forms not extending into the waters of the sea. The most salient example, at any rate the best known perhaps, is the genus *Pontodrilus,* the name of which fixes its habitat, and was naturally given to it on that account. It was originally found on the sea shore of the South of France, and I have myself examined examples from Nice. The worm lives among bunches of sea-weed cast up by the sea, and which are thoroughly salt. Besides the two forms that have been met with in this Mediterranean region but which are united by Dr Michaelsen into but one species, other *Pontodrilus* have been described from so many and such diverse parts of the world as the following. The West Indies (Bermudas, Jamaica etc.), the coasts of South America, of both West and Eastern Africa, the Red Sea, Christmas Island near Java, Sharks Bay in West Australia, the Hawai Archipelago, Celebes, South West Australia etc. In fact there is no great tract of the ocean excepting the antarctic region where this genus is not to be found. It is possible however that this latter statement is not correct and that New Zealand ought to be added. But the species described from those islands, viz. *Pontodrilus lacustris*, is not a marine form at all as its specific name denotes; nor is it quite certainly to be included in the genus. On the other hand a form from the Chatham Islands in the same quarter of the globe, described originally as *Pontodrilus chathamensis*, is to be referred to the antarctic region. Altogether some dozen species of *Pontodrilus* have been described by different naturalists; but quite recently Dr Michaelsen has reduced these to three only, which are *P. bermudensis* (F. E. B.), *P. litoralis* (Grube) and *P. matsushimensis* (Izuka), with the doubtful addition of *P. lacustris* already referred to. Whatever may be the ultimate verdict upon this question of species it is clear that the genus is widely spread upon the sea shores of the world and that forms from different regions show some fixed variations, which others may eventually agree with their original describers in regarding as definite species.

It cannot be said that any salient characters in the organisation of these worms mark them out from either terrestrial or fresh-water Oligochaeta. There are no such important variations of structure as can be seized upon to characterise them as inured to salt water. The genus agrees with many aquatic forms in the fact that the nephridia are not present in the earlier segments of the body, not indeed putting in an appearance until about the thirteenth segment or even later. They are thin delicate worms; but there is nothing distinguishing about this, while the feeble or absent gizzard is a character which is really difficult to correlate with habitat. Still we have here a whole genus which is marine in its habit. Among the Megadrili or earthworms there are not many other examples of these 'euryhaline' forms as they have been named. On the shores of Patagonia however and Kerguelen shore-living species of the mainly antarctic genus of earthworms *Notiodrilus* have been met with. And there are a few allied cases among the antarctic genera of Acanthodrilinae.

In addition to these terrestrial forms there are a few limicoline genera which are partly marine in their habit. Thus several species of the prevalently arctic and antarctic family Enchytraeidae are shore living. There are also marine Tubificids such as

Clitellio arenarius and *Tubifex ater* (not uncommon on British shores), marine Lumbriculids and a marine Naid from the Italian coast. These forms show no great differences from their fresh-water allies.

EARTHWORMS ORIGINALLY PURELY AQUATIC ANIMALS.

The very name Earthworm, so distinctive as it is of the habitat of these animals, seems to have been expressly invented in order to crystallise into one word the remarkable distributions of these creatures. They are with very few exceptions the most purely terrestrial animals that we know. There are a few Mammals like the mole and several underground Amphibians and Snakes in the tropics which share this habitat with the worms, probably because they chiefly prey upon them. But there is no group of animals that is characterised by a subterranean existence in the way that earthworms are. For we cannot put burrowing animals, such as the prairie dog and many rodents, into the same category. These make and seek their burrows for protective purposes and in order to bring forth their young in security. They do not feed beneath the surface of the ground or pass their entire lives in that situation. We have already in a previous chapter dealt with such exceptional forms of earthworms as do not lead an entirely subterranean existence; but as was pointed out in chapter I these exceptions are but few and the immense bulk of earthworms fully justify their name.

Nevertheless there are many arguments which tend to show that these purely land-dwellers have grown out of exclusively water-dwellers and even that the change from the one mode of life to the other has been accomplished comparatively recently. For there are here and there vestiges of structures which seem only fitted for an aquatic life; and in other cases structural changes have commenced which would appear to be in definite relation to the underground mode of life prevalent to-day. Let us consider for a moment the differences which obtain between the conditions of life in water or in soft mud at the bottom of pool or river, and those which are undergone by a dweller in stiff soil or vegetable débris. In the first case the medium is fluid or at most very soft, while the soil is at least stiffer and harder to traverse.

Secondly the transition between the very bottom of a pool and the top layers of the water is more or less gradual, while the stiff soil ends abruptly in the tenuity of the atmosphere.

A third point of difference is doubtless the smaller supply of readily available oxygen in the still pools and even rapid rivers, which in certain stagnant pools and in the bottom waters of deep lakes must produce a very vast difference in physiological conditions.

We have already dealt with the characteristics of the aquatic genera of earthworms, not only in detailing the general characters of the families which are found in this situation but also in studying the features which earthworms show in those cases where they have reverted to an aquatic mode of life. It remains in the present section to attempt to descry in the purely terrestrial forms the remnants of adaptations to an aquatic life which are no longer of service to them.

It is a noteworthy fact, that the continuous circle of setae which is met with in certain earthworms is by no means a character of such classificatory importance as it was at one time, perhaps, thought to be. It is true that we meet with this character in the genera *Megascolex* and *Pheretima* which are not very far from each other in the ystem and are at any rate members of the same sub-family, the Megascolecinae. But we also find the continuous circle of setae well developed in *Plagiochaeta* which is not so near to *Pheretima*, and an approach towards it in *Dinodrilus* and *Dinodriloides* which are equally remote perhaps from both *Pheretima* and *Megascolex* on the one hand and *Plagiochaeta* on the other. Still it may be urged that all of these genera are at least members of the family Megascolecidae and that the question of a character which thus merely shows affinity is not yet eliminated. It is therefore of particularly great importance that Dr Cognetti de Martiis should have met with the South American genus *Periscolex* which, undoubtedly a member of a totally distinct family, the Geoscolecidae, yet shows the same complete circle of setae. The reason for dwelling upon this particular anatomical character in the present connection is because it would seem to be a character specially suited to an underground life where there is an equal pressure all round the body and where progression would seem therefore to be best attained by a continual leverage round the circular body. And this view is strengthened by the sporadic occurrence of this modification in different families. We thus come to the conclusion that the opposite state of affairs is a remnant of an aquatic life, a conclusion which it is the object of the present section to discuss. More than this, it would seem that an equal disposition of the two bundles of setae on each side of the body was a less modified state of affairs than the restriction of the two bundles or pairs of setae to the ventral surface, such as occurs for example in the genus *Dichogaster* and which is very obvious in some of the larger-sized members of this extensive genus. For the restriction of the setae to the ventral surface obviously favours progression upon a surface and not through a medium. And it is only among the terrestrial Oligochaeta that this mode of progression occurs. It might also be urged, and with some reason, that the retention of rather longer setae upon the clitellum in the Lumbricidae and Geoscolecidae, and the possession of equally long or in many cases much longer setae corresponding to one of the two pairs of setae of the generative segment in certain Megascolecidae, is a feature in which an aquatic condition—so to speak—is retained. The setae would represent a vestige of the general presence of long setae over the body generally such as is convenient or at least not inconvenient to an Annelid living in water or soft mud. But probably it will be thought the modified genital setae are a recent development and not a retention.

There is no more thoroughly terrestrial family of earthworms than that of the Moniligastridae and yet this family in its general anatomical characters shows many points of likeness to aquatic forms as has been now pointed out by many observers. It is true that these characters are not those which might be associated at first sight with an aquatic life. But none the less they are characteristic of most of the families which live in the waters of the earth. Thus *Moniligaster* and its allies (*Eupolygaster*, *Drawida*, etc.) have quite short sperm ducts which open on to the exterior at furthest in the segment next to that in which their internal funnel lies. Again the simple structure of the terminal gland into which they open and which in its turn opens on to the exterior is very like that of such a family as that of the Lumbriculidae. Another fact is the simple undivided cavity of the sperm sacs which is unlike that of typical earthworms but again like that of all of the Limicolous families. We may fairly see in these worms evidence of origin from aquatic ancestors. Evidence of the same nature, *i.e.* not as showing the retention more or less of anatomical characters commonly associated with a life in water, but as affording indirect evidence of an origin from actually aquatic forms, is to be seen in certain members of the families Geoscolecidae and Eudrilidae. In both of these it not infrequently happens that the sperm sacs are but a single pair and that that pair consists of sacs of extraordinary length. Thus in *Trichochaeta* (or ***Hesperoscolex***) *barbadensis* Miss Fedarb and I have shown that the long thin sperm sacs extend through no less than 109 segments, which is vastly in excess of the length of those of the majority of earthworms in which they are most commonly limited to a single segment. In the same way the Eudrilid worm *Polytoreutus magilensis* has a pair of long and thin sperm sacs which extend through some fifty segments. This elongation of the sperm sacs in the ripe worms is a very common feature of the Limicolous genera.

CHAPTER III

THE EXTERNAL FEATURES OF EARTHWORMS AND THEIR RELATION TO HABIT AND ENVIRONMENT

To the very inexperienced eye all earthworms might appear to be quite similar in detail as they undoubtedly are in general form. But it needs not a great deal of examination to detect even salient characteristics whereby one kind may be distinguished from another; to the expert it is possible in very many cases to go no further than the outside before assigning its correct place in the system to a given example. The general external features of this group of worms have been already dealt with in another chapter. To some of these we again direct attention in a more elaborate fashion in order to emphasise the possible meanings of the variations met with apart from their use in systematic arrangement. It is difficult to say in comparing one worm with another what is the most salient external difference. There are however a few which may be regarded as equally conspicuous on a nearer examination of the specimens. The varying position and greater or less extent of the clitellum, the longer or shorter retractile or non-retractile prostomium, the position of the usually conspicuous male pores, and the existence of in the first place and—when present—the numbers and situation of the so-called genital papillae are among the most obvious. The setae and their position we treat of under the heading of the modification of the worms to lead a terrestrial life; and though these chitinous organs differ greatly they do not concern us in the present section. The girdle or clitellum ('eminentia quasi ulcerata') has been long observed as a character of these animals and it is one which distinguishes them from all other worms except the leeches and a very few marine Polychaeta. This modified region of the body is often of a different colour to the rest and has a glandular look which readily enables one to recognise its position and limits, though its obviousness is less in some cases. It either forms a complete ring round the body or is developed upon the dorsal surface and only to a slight extent upon the ventral surface. Its use, as is well known, is to secrete the cocoon in which the eggs are deposited; and the epidermis which forms it is thickened and more glandular than that in other regions of the body. Among earthworms it is doubtful whether the clitellum ever occupies less than three segments; it consists of three only in the great majority of species of the marked genus *Pheretima*. From this lowest level it extends in other forms, and in the partially aquatic African genus *Alma* it may occupy as many as forty segments. The position also varies from genus to genus and from species to species. It is sometimes further forward and sometimes further back. In the remarkable family Moniligastridae this organ is developed earlier in the body than in any other group of true earthworms, consisting of four segments or so commencing with the tenth. As a rule the clitellum begins further back than this—the thirteenth or fourteenth being a common place for the first commencement of the organ among the Megascolecidae, while among the Geoscolecidae and Lumbricidae it is generally much further back, commencing in *Alma* at the forty-fifth. These details might be increased to many pages; but enough has been said to emphasise the variability of the organ. What reason can be assigned to this variability, which might be supposed unnecessary in view of its functions? There are perhaps two suggestions that may be made, though many facts are lacking which might offer confirmation or refutation of either of these. It is to be noticed that on the whole the older types such as the Moniligastridae and the Megascolecidae (including for this purpose the Eudrilidae) have clitella which are short. There are a few but not many exceptions. These older types do not seem capable of extending their range with any rapidity. It is true that here again there are exceptions, notably many species of *Pheretima* which are considered under the section which deals with the migration of these animals. On the other hand the Lumbricidae have on the whole a more extensive clitellum and so have many Geoscolecidae. It is obvious that of all earthworms the Lumbricidae is the family which has the greatest capacity of migration and adaptation to new circumstances. The reason for this may be that in the latter case the more extensive clitellum produces a larger cocoon which in its turn can hold and cherish while they reach maturity a larger number of embryos. Much remains to be learnt under this heading. But the comparatively small clitellum of the large Ceylon *Megascolex coeruleus* only contains two embryos, while the also comparatively small cocoon of the large and restricted *Octochaetus multiporus* (limited to the South Island of New Zealand) only contains a single embryo. This latter fact may be regarded as fairly well established since I myself examined quite fifty cocoons.

On the other hand larger numbers seem to arrive at maturity in the cocoons of *Allolobophora*. The more extensive clitellum must produce a relatively larger cocoon, and it is interesting to note that the cocoon of the widely distributed genus *Criodrilus* (Europe and South America) is very long although not of great diameter. However the facts are not sufficiently great to dogmatise much upon this subject. Another conceivable reason for differences

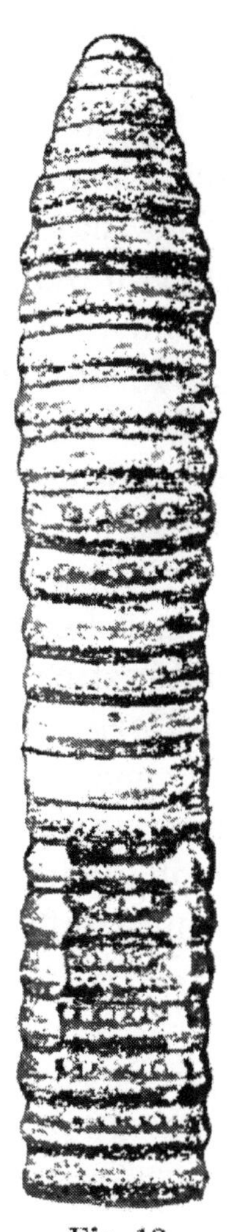

Fig. 12. Fig. 13.

Fig. 12. Ventral view of *Pheretima solomonis* to show papillae which are to be compared with those of fig. 13. (×2.)

Fig. 13. Ventral view of *Pheretima sedgwickii*. (×2.)

in the clitellum is—as I also think is the case with the genital papillae—to prevent hybridisation. That the sense of touch is delicate in these animals seems clear from the abundant development of epidermal sense-organs. It may be that the feel of the clitellum during union enables two individuals of a given species to come together and prevents those of different species from mating. In any case there is no positive evidence that hybridisation does occur in this group of animals. The astounding variability and yet constancy in a given species of the genital papillae is in favour of regarding these organs as tactile recognition marks; and it will be noted that they are not by any means characteristic of some of the older types of earthworms. Furthermore they are particularly conspicuous in such genera as *Pheretima*, *Megascolex* etc., which possess a large number of species. In these of course mutual recognition would otherwise be more difficult.

CHAPTER IV

SENSE ORGANS AND SENSES OF EARTHWORMS

As this is not an anatomical treatise we shall not attempt any detailed anatomical and histological account of the sense organs in this group of worms.

But a few facts must be given in illustration and explanation of the senses of touch and sight that the Oligochaeta undoubtedly possess. These Annelids, unlike their allies the marine Polychaeta, and even their allies on two other sides, the leeches and flat-worms, have no complexly organised eyes or other sense organs. They have in fact no organs to which a definite sense can be attached on histological grounds. There is nothing comparable to an eye or to the auditory sacs of other low worms. There are only particular cells of the epidermis, often associated into small groups, and those again into larger associations of rows of such groups of cells. It is to be presumed that these modified groups of cells have a sense function; but no more can be said than that they are doubtless tactile and also to some extent receptive of the influence of light. True visual cells have been asserted to exist in earthworms, consisting of cells of which a part is clear and transparent and has been supposed to serve as a lens for the rest of the cell which represents a retina. But belief in the function of these cells is by no means unanimous. On the other hand many investigations have proved the existence of groups of epidermic cells of an oval form which are at present arranged in definite rows upon the segments of both terrestrial and aquatic forms, which are moreover connected with nerve terminations, or are at least—according to more modern views—in close contact with the terminations of nerve fibres. These are furnished often at their free tips with minute sensitive processes. There is nothing in the structure of these to associate them definitely with any sense in particular. They suggest of course tactile organs more than organs of any other sense. In addition to these are certain problematical organs which are found in the Eudrilidae and are present in the members of one section of that group, the section which is represented by the universally found *Eudrilus*.

These bodies have been compared to a Pacinian body (a sense-organ found in Vertebrates) and they bear no little resemblance to it. For each consists of a darkly staining core surrounded by a layer of cells arranged like the coats of an onion. In any case it would appear that these bodies must be looked upon as of a sensory nature, though they do not reach the surface of the body but underlie the epidermis. Their function must remain purely a matter of guesswork at present, for nothing to the point is known of the habits of the Eudrilidae. It has been suggested by Dr Gustav Eisen that these cells are auditory and serve to warn the worm of the footsteps of birds and other enemies. That too is his view of certain peculiar but different cells found in the epidermis of *Pontoscolex*. In the latter something like an otosome has been found which is certainly lacking in the Eudrilidae, whatever may be the function of the cellular epidermic bodies here briefly referred to.

While there is thus nothing very positive to be gleaned from an examination of the structure of the Oligochaeta as to the senses which they may possess, experiment has done something towards an elucidation of their behaviour under stimuli and their reaction to light and to other forces which come into play during their lives. There is some evidence that earthworms can see, using that expression of course in a very broad way. At any rate they react to changes in the intensity of light. The gross experiment of flashing a lantern upon earthworms which are reaching out from their burrows with the tail end inserted in those burrows shows that they have an appreciation of light. More refined experiments have been conducted upon the sense of light. Dr Graber used a box with two compartments, the one of which was dark and the other illuminated with diffused daylight, *i.e.* not direct sunlight. The worms were allowed equally free access to both and were examined at the close of every hour, and their positions noted. The investigator found that on the average the dark half of the box contained 5·2 times as many worms as the light chamber, thus indicating a very marked preference for absence of light.

Not only is this the case, but the same observer

proved that earthworms can distinguish between degrees of intensity of light. This obviously indicates a more complete 'visual' sense. He illuminated the light-box of the former experiment with light admitted through a ground glass screen, thus diminishing its intensity. The other chamber was left as before but the screen was removed, thus admitting full daylight. In this experiment the number of worms in each compartment proved to be different. The results were not so striking as before, since only rather more than one-half were found in the more dimly illuminated chamber. It is a well-known fact that if earthworms are abroad at all from their burrows, it is during the night that this movement takes place, the numbers decreasing towards morning though worms are often seen crawling about well after sunrise. Some experiments have been made in attempt of explanation of this apparent anomaly. It would appear from these experiments that while worms are negatively phototropic to strong and moderate light as has already been pointed out, they are positively phototropic to very dim light; hence the advent of evening calls them forth from their burrows. It will be noted that this perception is of very great advantage to the worm since its more active enemies above ground are diurnal. It was held originally that the head end of the worm only was thus sensitive to light; but more recent experiments have shown that this is not the case, and that all of the body is sensitive. This disposes of course of the existence of special light-receiving organs in the anterior part of the worm's body. Not only this, but an interesting extension of the view has been promulgated. It has been shown by Prof. G. Parker and a colleague that in the common Brandling worm, *Allolobophora foetida*, the response to light stimuli was related to the amount of the body exposed to its influence. This is very important as showing that the light perception in these creatures is probably not due to special organs having a limited position on the body, but is due to collective sense impressions of many cells scattered over the whole body, the impression being the greater when the whole body is exposed and less when only parts of it are exposed. Furthermore, and this has quite another importance, these observers noted that the reaction effects differed when only a part of the body was exposed; that they were greater in the front of the body, less in the middle, and less still at the tail end. Indeed they found that the reactions in the case of the front end of the body alone being exposed were rather more than one-third as compared with those which were shown when the whole body was subjected to the light stimulus. The fact that the least sensitive region of the body is the posterior end has, it is true, only been definitely proved in the case of the worm whose specific name has been mentioned. But it is possible that others are similarly affected. And it is highly important to note the prevalent habit among the Tubificidae of lying with the head end imbedded in the mud of the pool which they inhabit, while the tail end emerges and waves freely in the flood. The additional fact that this tail end occasionally bears gills (as in *Branchiura sowerbii* and *Phreodrilus branchiatus*) has a collateral importance not to be mistaken.

CHAPTER V

RELATIVE FREQUENCY OF EARTHWORMS IN DIFFERENT REGIONS OF THE WORLD

It will be of use for various purposes to be considered later to arrive at a comprehensive view of the relative numbers of species and genera of earthworms in the four quarters of the globe. And in making this general census we shall not take into consideration the purely aquatic forms, but shall limit ourselves to the earthworms, *sensu stricto*, or Megadrili, of which, however, it is true that some members are actually lake and river dwellers. This latter fact will not, however, interfere with the usefulness of the comparative survey.

Two preliminary remarks are necessary. The opinions of naturalists vary as to the limits of genera; and a species may be a species to one and a mere variety to another. Thus it will be impossible to give a summary of the facts to be enumerated presently, which will be either absolutely accurate or which will satisfy everyone in every detail. But it is asserted that the following survey is substantially correct.

In the second place it is often possible to eliminate from the fauna of a given region those species and even genera which have been accidentally imported, a matter which will receive careful consideration on a later page. Such forms are therefore, in those cases at any rate where the evidence seems to be overwhelming, withdrawn from the list. In other cases, particularly in the Eastern region of the world, it has been found less easy to rectify the catalogues by removing what Dr Michaelsen has termed 'peregrine' forms.

We shall commence with a census of South America; the entire Continent will be divided for the present purpose into three divisions, viz. South America, Central and North America, and in the third place the West Indian Islands.

In South America we find that the bulk of the indigenous earthworms belong to the family Geoscolecidae and to a definite sub-family, viz. Geoscolecinae.

These genera are *Onychochaeta* with one species, *Hesperoscolex* of which one species is known from the area, *Periscolex* with one species, *Anteoides* with two species, *Pontoscolex* one species. *Opisthodrilus* with two species, *Rhinodrilus* (including either as synonyms or as sub-genera, *Thamnodrilus, Anteus, Tykonus, Urobenus* and *Aptodrilus*) with no less than 49 species: *Andiodrilus* with five species, *Holoscolex* with one species, *Glossoscolex* ten species, *Fimoscolex, Andiorrhinus* and *Enantiodrilus* with one species apiece.

Thus of this sub-family of Geoscolecidae we have in South America a large number of genera and a much larger number of species. Of a second sub-family of Geoscolecidae there are three species of *Criodrilus* found in the South American Continent.

We now turn to the Megascolecidae of which a large number of species occur within the area now under consideration. The bulk of these belong to the sub-family Acanthodrilinae and they are as follows:

Of the genus *Notiodrilus* there are ten species, of *Microscolex* two species, of *Chilota* 19, of *Yagansia* 13.

A second sub-family Trigastrinae also occurs in this Continent and the following genera are permanent inhabitants, viz.:—

Dichogaster (=*Benhamia*) with only three species,

of which two at least are found elsewhere, and of which therefore the autochthonism is doubtful.

Finally, there is the sub-family Ocnerodrilinae comprising the following genera: *Kerria* with ten species, *Ocnerodrilus* (with sub-genera *Liodrilus*, *Ilyogenia* and *Haplodrilus*) four species, which again are rather doubtful indigenes of the South American Continent.

Leaving aside certain species (of the genera *Lumbricus*, *Pheretima*, etc.) which are clearly not indigenous, the South American Continent harbours 150 kinds of earthworms which are distributed in some 19 genera. But of these a few species (*e.g.* *Onychochaeta windlei*, *Kerria macdonaldi*) stray into neighbouring regions, *i.e.* the West Indies and California. It is doubtful therefore whether they are to be referred to as limited to one of these regions and accidentally imported into the others, or whether they are genuine inhabitants of both.

The South American Continent shares with the West Indies the following genera, but the species (except in the case of *Onychochaeta windlei*, *Glossoscolex peregrinus*) are distinct; these genera are *Hesperoscolex*, *Pontoscolex*, *Dichogaster*, and *Ocnerodrilus*. *Diachaeta* is limited to the West Indies.

The following South American genera are also found in Central and warmer North America (Mexico, California), viz. *Hesperoscolex*, *Periscolex*,

Rhinodrilus, Pontoscolex, Andiodrilus, Glossoscolex, Notiodrilus, Microscolex, Dichogaster, Kerria and *Ocnerodrilus*. But with the exception of *Hesperoscolex*, which seems to belong rather to Central America and the West Indies, *Microscolex* and *Pontoscolex* which are world-wide and whose original home is therefore difficult to fix, and *Dichogaster* and *Ocnerodrilus* which would seem rather to be rare immigrants (perhaps not truly indigenous) into South America, these genera are practically distinctively South American.

Thus we may fairly say that the genera *Anteoides, Opisthodrilus, Andiodrilus, Holoscolex, Glossoscolex, Rhinodrilus, Andiorrhinus, Fimoscolex, Enantiodrilus, Notiodrilus, Chilota, Yagansia* and *Kerria* are at least very distinctive of S. America and that they are represented by the large majority of species found in that continent, the total being 120 or slightly more.

Leaving the American Continent and adjacent islands and archipelagos, the next great land-mass to receive attention will be the Continent of Africa. In giving a census of the species and genera of earthworms which inhabit this quarter of the globe it must be premised that the facts relate only to Africa south of the Sahara. But little is known of the genera which occur in Algeria and Morocco, but from what little is known it is clear that they should come into consideration in connection with the fauna of Europe and not with that of tropical and south temperate Africa.

We have in the first place to consider an entirely peculiar family, of fair extent in genera and species, which is limited to this region of the world; that is to say with one apparent exception which is clearly only apparent. The genus *Eudrilus* is one of the few kinds of worms that turns up in collections from every tropical part of the world and even at times from more temperate countries. It is one of those 'peregrine' forms, as Dr Michaelsen has termed them, which possess unusual facilities for extending their range. Presumably its real home is Africa. This family is known as the Eudrilidae though by some it is only regarded as a sub-family of the Megascolecidae. In this family we have the following genera: *Eudriloides* with 11 species, *Platydrilus* with 11 species, *Megachaetina* with two species, *Reithrodrilus* with one, *Stuhlmannia* with five species, *Notykus* and *Bogertia* with one species each, *Metadrilus* also with but one species, *Pareudrilus* with perhaps five, *Libyodrilus* with one and *Nemertodrilus* with two species, *Metschaina* with two species, *Eudrilus* with two (or possibly more?) species, *Parascolex* with four species, *Preussiella* with two, *Buttneriodrilus* with two, *Eminoscolex* with 16 species, *Hyperiodrilus*, *Heliodrilus*, *Alvania*, *Rosadrilus*, *Kaffania*,

Euscolex, Metascolex, Beddardiella, Gardullaria with only one species to each genus, *Bettonia* with three species, *Teleudrilus* with 15 species, *Polytoreutus* with 22 species, *Iridodrilus* with two, *Malodrilus* also with two, *Neumanniella* with eight, *Eupolytoreutus* with two and *Teleutoreutus* with only one species.

This completes the list of the Eudrilidae. We will take the huge family Megascolecidae next; and we find in tropical Africa that the genus *Dichogaster* alone contains at least 93 species confined to Africa, as well as a few which it shares in common with America, and the common and widely spread *D. bolavi* which has even made its way to Europe. Of the sub-family Ocnerodrilinae we have the genus *Gordiodrilus* with seven species, a genus which also occurs in Madagascar and the West Indies: *Nannodrilus* with three species, *Diaphorodrilus* with one species and a few examples of *Ocnerodrilus* and its sub-genera, some of which are also forms that occur elsewhere in the world, for example *Nematogenia panamensis*. Of *Pygmaeodrilus* there are eight or nine species.

Of the sub-family Acanthodrilinae the Cape region of South Africa harbours some seven species of the genus *Notiodrilus*, of which one however is a West African form. The allied *Chilota* has 13 species, and there is a peculiar genus *Holoscolex* near to *Yagansia* with one species. We next have to deal with the Geoscolecidae, of which a sub-family, Microchaetinae, is mainly found in Africa, the rest being found in the neighbouring Madagascar and some few in the East. *Microchaetus* contains about 14 species, *Tritogenia* perhaps three, *Callidrilus* two, and the genus *Glyphidrilus*, mainly found in Asia, has one species in the region now under consideration. In addition to these Geoscolecids there is the peculiar and largely aquatic *Alma* with six or seven species in East, West, and Central, Africa, and in the Nile region.

Summing up the genera which are found in tropical and South Africa we find that there are 44 which are abundant in, or entirely confined to, the region. In addition to these four or so occur in Africa but are either more abundant elsewhere or (as in the case of *Chilota* and *Notiodrilus*) are equally distinctive of other parts of the world. The number of species may be estimated at 270, possibly rather more. Clearly therefore this part of the world is much richer than South America, both in numbers of genera actually found, and peculiar to the country, and numbers of species.

Passing from Africa the next definite quarter of the globe which will detain us here is the island of Madagascar, so remarkable for the Mammalian fauna which characterises it, for its lemurs, peculiar Insectivora and Carnivora, and above all by reason of the absence of the prevalent African types such as antelopes, zebras, rhinoceros etc. It is probable that a good deal still remains for discovery among the earthworms of this island; but a considerable number are already known and they are as follows:

The Eudrilidae are completely absent, a state of affairs which is paralleled by the absence of antelopes among mammals.

The sub-family Acanthodrilinae of the family Megascolecidae are represented by four species of the genus *Notiodrilus*, with which perhaps *Maheina braueri* from the Seychelles should be included as it presents small differences from *Notiodrilus*.

Among the Megascolecinae a good many species of *Pheretima* have been collected both on Madagascar and on certain adjacent islands; but these, with one possible exception, are forms which occur elsewhere and are often indeed very widely distributed 'peregrines,' so that it is hardly permissible to place them among the real inhabitants of Madagascar. The same arguments hold in the case of *Lampito mauritii* and the ubiquitous *Eudrilus* and *Pontoscolex*. But there is the peculiar *Howascolex*.

Among the Ocnerodrilinae a distinct species of *Gordiodrilus* occurs, and an obviously introduced *Ocnerodrilus*.

It is among the Geoscolecidae that the most characteristic forms are met with. These belong exclusively to a genus of the Microchaetinae, *Kynotus*, which is found nowhere else but in Madagascar, where it is represented by at least twelve different species.

We have therefore in Madagascar and the surrounding islands only two peculiar genera, only four genera which can be regarded as endemic, and only about seventeen peculiar species.

Passing eastward and crossing the Indian Ocean we come to the Continent of Asia, and we shall first direct attention to the peninsula of India and adjacent parts of Burmah and the island of Ceylon, of which there has been accumulated a great deal of knowledge concerning the Oligochaetous fauna.

This quarter of the globe differs quite as much from any that have been hitherto considered as they do among themselves. We have left behind us the Geoscolecidae with the exception of the ubiquitous and clearly peregrine *Pontoscolex*, and the genus *Glyphidrilus*, which, being aquatic at times, is perhaps hardly to be considered in the present survey.

We have also in this Indian region the equally ubiquitous *Eudrilus eugeniae* which need not of course detain us further. The Lumbricidae are for the most part of European forms with the exception of *Helodrilus indicus* conceivably an actual inhabitant of India. It is among the Megascolecidae that the vast majority of the forms endemic in India are to be found. We shall take this family according to its sub-families. In the first place we note that the sub-family Acanthodrilinae is totally unrepresented. The large sub-family Megascolecinae has very numerous representatives. Of the genus *Megascolex* itself there are some 30 species, of which, however, two or three are looked upon as varieties. Of the allied genus *Notoscolex* there are ten species and of *Perionyx* about 13. The genera *Megascolides, Diporochaeta, Spenceriella* and *Woodwardia* have only six species between them of which three belong to *Megascolides. Plutellus* has five species in this region. *Lampito,* which is a widely spread form with but one species *L. mauritii,* may or may not find here its original home. It also occurs in Africa and Madagascar. *Pheretima* is represented by no less than 12 species, of which at any rate the very great majority are peregrine forms, and not to be safely regarded as forming an integral part of the fauna of the Indian peninsula and adjoining countries.

Of the sub-family Octochaetinae the type genus *Octochaetus* has ten species in this region and *Eutyphoeus,* which is restricted to the region, has 15 species or perhaps rather more. *Hoplochaetella* has but one. Of the Trigastrinae there are four or five examples of the genus *Dichogaster* which occur within the region now under consideration; but it is doubtful whether they are truly indigenous. On the other hand *Eudichogaster,* closely allied both to *Dichogaster* and *Trigaster,* seems to be confined to this part of the world where it is represented by five species. Finally we come to the last sub-family—that of the Ocnerodrilinae, which is represented by a species apiece of the genera *Ocnerodrilus*, *Nematogenia* and *Gordiodrilus*. The latter species *G. travancorensis* is alone to be regarded as endemic and it is very near to the African *G. zanzibaricus*, described some years since by the present author.

The remaining family of terricolous Oligochaeta found in India is the family Moniligastridae which is practically limited to this part of the world and consists of at least twenty species distributed among the genera *Moniligaster*, *Eupolygaster*, *Desmogaster* and *Drawida*, the majority belonging to the last-named genus.

This quarter of the globe is therefore inhabited by 18 genera which are certainly truly endemic, and which comprise between them about 120 species. But only four or five genera are peculiar.

The remainder of the Asiatic Continent is not very well explored with regard to its earthworm inhabitants. It seems clear however that the southern and coastal region of China and Japan with the Malay peninsula are really continuous with the mass of islands which lie between India and Australia and form together a tract of land which is characterised by an Oligochaetous fauna differing from both that of India on the one hand and Australia on the other.

We shall therefore consider this huge portion of the globe as one region comparable to the other divisions that have been hitherto considered. With reference to the Lumbricidae and Geoscolecidae the same remarks may be made as in the case of India. The indigenous forms of the latter family are to be looked upon as outside of the present survey since they are largely or entirely aquatic forms. *Pontoscolex corethrurus*, and *Eudrilus eugeniae*, need not detain us for reasons already amply stated. We now come to the great family Megascolecidae. Of this family the genus *Pheretima* stands at the head; and of the 200 or so species that have been or can be assigned to this genus all, with merely two or three exceptions, are natives of the Eastern Archipelago and adjoining mainlands of Asia. Of other Megascolecinae the region has yielded the following genera. In Java one species of *Woodwardia* (*W. javanica*) has lately been described. It is regarded by Michaelsen however as doubtfully indigenous. And the same remark may be made of *Perionyx*. *Plionogaster*, however, with four or five species, is limited, as far as our present knowledge goes, to the Philippines and to neighbouring islands. The Acanthodrilinae and Octochaetinae are totally absent from this part of the world, there being no record even of peregrine species of these sub-families. The sub-family Trigastrinae is not however unrepresented; for of *Dichogaster* several species occur, such as *D. malayana* and *D. saliens*. But these are by no means certainly to be looked upon as real natives of the situations within this area where they occur. This completes the scanty list of genera found in the region under consideration; for one Moniligastrid (*M. barwelli*) is hardly to be looked upon as indigenous. We have therefore to record here but five genera, of which only two are certainly indigenous and probably also confined to the region; they contain between them certainly two hundred species.

We next come to the Continent of Australia. The earthworm fauna is again quite without Geoscolecidae and of course Lumbricidae. As to the former there are not even doubtful cases like *Glyphidrilus* of the East; for we find only recorded *Pontoscolex corethrurus*, a species concerning whose extraordinary powers of migration there is no possible doubt. The same may be said of *Eudrilus* also recorded from Australia. Here the Megascolecidae are as conspicuous as in the old world generally. We find, however, a great many members of the sub-family Acanthodrilinae. Of the antarctic genus *Notiodrilus* there are some six species. *Microscolex* occurs; but the real habitat of this genus is very doubtful. *Diplotrema*, with one species, *D. fragilis*, is not only indigenous to, but confined to, Australia.

The most prevalent sub-family is that of the

Megascolecinae. Of the genus *Pheretima* there are two species which may or may not be truly indigenous. Of *Plutellus* there are forty species; there are over thirty of *Diporochaeta*, while *Notoscolex* (with which Dr Michaelsen associates *Digaster*, *Didymogaster*, *Perissogaster*) includes more than forty species, *Fletcherodrilus* has but one species, *Megascolex* has seventy species, *Woodwardia* sixteen, *Spenceriella* five, and *Megascolides* seventeen. There are four species of *Perionyx* which were at one time regarded by Michaelsen as necessitating a new but allied genus *Perionychella*; the two are now merged. Thus there are not far short of 150 species and eleven genera represented, of which only two are limited to Australia.

Having completed the survey of the central and southern land masses of the globe we next direct attention to the northern land masses, viz. North and Central America on the one hand, and Europe and northern Asia on the other. With regard to America we find besides many species of Lumbricidae the genera *Notiodrilus*, *Microscolex* among the Acanthodrilinae, *Megascolides* and *Plutellus* of the Megascolecinae, no member at all of the Octochaetinae, a considerable number of species of *Ocnerodrilus* and sub-genera belonging to the Ocnerodrilinae, a good many species of *Dichogaster* and at least one of *Trigaster* among the Trigastrinae, while one sub-family, that of the Diplocardiinae, is only found here and contains two genera *Diplocardia* and *Zapotecia* with quite ten species between them of which only one belongs to the last-named genus. There are no Geoscolecids (except the chiefly aquatic *Sparganophilus*); this family stops short in the West Indies where their presence has been briefly referred to in considering the worms of South America. The vast majority of the genera enumerated here are only found in the warmer parts of the North American continent. We have therefore in this division of the world some nine genera of which at least one, viz. *Diplocardia* (and its close ally, hardly perhaps to be separated, *Zapotecia*), is confined to it, while *Trigaster* only extends as far south as to the West Indies.

Europe and northern Asia, of which Europe alone and certain limited tracts of Asia are at all known, contain all the genera of the family Lumbricidae which are (according to Michaelsen) *Eiseniella*, *Eisenia*, *Helodrilus*, *Octolasium*, *Lumbricus* with various sub-genera of *Helodrilus*, such as *Dendrobaëna* and *Allolobophora*. Besides these forms, which amount to at least 130 species, we find the genus *Hormogaster*, with two species, the only genus of the Geoscolecid sub-family Hormogastrinae; there is also a species apiece of the genera *Criodrilus* and *Sparganophilus* (sub-family Criodrilinae). But as these are at least largely aquatic they come under another set of rules and are not important in the present survey of the earthworms of the world. It will be thus seen that with these two last exceptions the area in question contains but six genera of which *all* are peculiar.

We have now completed the survey of the principal land-masses of the globe. We shall conclude with a reference to one of the largest islands of the world, viz. New Zealand, with which will be included a few outlying islands such as the Aucklands, Snares Island, etc. The reason for not ignoring these islands as we have ignored certain other islands of large size, such as Japan, will be apparent from the peculiarity of the earthworm fauna which they show. From New Zealand the Moniligastridae, Geoscolecidae, Lumbricidae, and, of course, the Eudrilidae, are absent, save the ubiquitous *Eudrilus*. The only family of earthworms which is here represented is that of the Megascolecidae. Of this family the Acanthodrilinae are very well represented. We have at any rate seven species of the genus *Notiodrilus* of which the bulk are from the small adjacent islands and not from the mainland. The genus *Rhododrilus* with nine species is actually limited to the New Zealand group and so is an allied genus consisting of two species only, viz. *Leptodrilus*. This latter genus is confined, so far as present information goes, to the Auckland and Campbell islands. *Dinodriloides* with two species is also limited to New Zealand and to the North Island. *Maoridrilus* with some ten species is another native and restricted genus. So too is the allied *Plagiochaeta* with numerous setae on each segment but with the alternating and single nephridial pores of *Maoridrilus*. There are several species of *Plagiochaeta* of which one has been lately regarded by Dr Michaelsen as really falling within the otherwise Indian genus *Hoplochaetella*, while for another he has formed the genus *Pereiodrilus*. *Neodrilus* with but one species *N. monocystis* is another peculiar New Zealand genus. The family Octochaetinae contains only four genera, of which one, viz. *Dinodrilus* (with four species), is limited to New Zealand, while *Octochaetus* has about five representatives. *Hoplochaetella* (if Dr Michaelsen's surmise referred to above be correct) has one species in New Zealand. The Megascolecinae are less numerous than the Acanthodrilinae, but there are ten species of the genus *Megascolides* (which includes Benham's genus *Tokea*), perhaps seven species of *Diporochaeta*, and two other species which Michaelsen has removed from the genus *Diporochaeta* and placed in *Spenceriella*. One *Plutellus* (which however may have been introduced) completes the New Zealand Megascolecines. We have therefore in this part of the world fifteen genera including between them some 58 species; eight of the genera are peculiar to the islands.

From this brief statement of facts some inferences of interest can be drawn. It is in the first place plain that every part of the world except the extreme north and south has a considerable fauna of earthworms. The one exception would appear to be the northern part of the North American continent. Here we meet with members of the family Lumbricidae which are however species that are met with in the Euro-Asiatic province and are thus to be regarded as possibly later immigrants introduced probably by man. Thus temperature short of a constantly frozen condition of the ground is not a bar to the existence of earthworms. Even a freezing of the ground for lengthy periods is not a complete obstacle to the existence of those Annelids; for I have myself received examples of Lumbricidae from the arctic island of Kolguev. Moreover the temperate regions would seem to be as fully populated in the way of individuals, and even of species, as are the tropical regions. Indeed as to individuals it seems that the temperate regions are more fully supplied than much of the tropics. This however is not quite the object of the present section to discuss. We are here concerned with the relative frequency of genera and species. There are according to a recent estimate of the Rev. H. Friend some forty species recognisable in Great Britain. And as already has been stated the earthworms of Europe amount to perhaps 130,—at any rate well over one hundred. In tropical America there are hardly more. But in the latter case the number of genera is very greatly in excess of that of Europe. We cannot however say that an abundance of generic types is quite characteristic of the tropics. For the Eastern Archipelago, though rich in species, is but poor in genera, not possessing more than half a dozen or so. And on the other hand the temperate climate of New Zealand has produced a very considerable series of genera, much more than those of the islands of the East and nearly as many as those of, for instance, Central America and the West Indies.

This conclusion is in its turn contradicted by the conditions observable in Chili and the temperate regions of South America, where the number of species is large but the number of genera small. In short no general laws, in the present state of our knowledge, can be laid down as to the connection between species and genera on the one hand and climatic conditions on the other. In this department of our subject we cannot do more than has already been done, *i.e.* to state the actual facts. One is tempted in comparing the rich fauna of tropical Africa with the very limited fauna of Madagascar to associate a richness of types with extent of land surface. In the two cases cited this conclusion is obvious. It may also be extended—if we confine ourselves to species and not to genera. For the two great islands of New Zealand have not between them more than fifty species of earthworms, while Australia has four or five times that number. It will be noticed however that we cannot associate poverty of generic differentiation with limited land masses; for New Zealand has a large number of generic types, very many more than the huge Euro-Asiatic tract of continent.

THE RANGE OF GENERA.

We have seen, and shall again refer to the fact, that individual species of earthworms have not as a rule a range over a great extent of country, save only in those cases such as *Pheretima heterochaeta* which belong to that physiological section of these worms called 'peregrine' forms; these appear to possess some means of extending their range by the assistance of man which is denied to other forms. Apart from these instances, which do not come under the present category, it is only *Lumbricus* and its immediate allies, *Helodrilus*, etc., of which certain species are found to exist over wide tracts of land. There are however many genera which have a wide range and which may be contrasted with others in which the range is very limited. The two extremes are moreover connected by forms with an intermediate range. There is no doubt whatever that the genus with the widest range is *Notiodrilus* of which species are found throughout the antarctic region, viz., in Patagonia, the islands of the Antarctic Ocean, the Cape of Good Hope, New Zealand, and also further to the north, sometimes even to and beyond the tropics in America, Australia, and Africa. There is no other genus of which the genuine extension (*i.e.* not in any way due to man) is so great as this genus *Notiodrilus*. And this fact gains much significance from the now generally accepted view that in its anatomical structure *Notiodrilus* comes near to the original type of earthworm.

Perhaps the next most widely distributed genus is *Helodrilus* of the family Lumbricidae which occupies Europe and Asia to the extreme east, and is thought also to be indigenous to certain parts of North America. But this range, though equally wide perhaps in mileage, is less impressive than that of *Notiodrilus*, since the land areas inhabited by the genus are continuous—almost so if we accept North America as its real habitat. Here we have a case precisely the opposite of that of *Notiodrilus*; for while there are reasons for regarding *Notiodrilus* as an ancient form of Lumbricid, there are equally good reasons for regarding the Lumbricidae as the most modern family of earthworms.

To find other instances of widely spread genera we must recur to the great family Megascolecidae. There are practically no Geoscolecidae which have a really extensive range. The only instances are *Criodrilus* and its ally *Sparganophilus* which occur in America, whether North or South, and in Europe; but as these forms are at least largely aquatic the facts are not quite comparable with those now under consideration.

The genus *Dichogaster* (which includes as synonyms *Benhamia, Millsonia, Microdrilus*) is unquestionably indigenous to tropical Africa and certain parts of America including the West Indies. It has been also met with in the East; but as the species there occurring, such as for example the species originally described by myself as *Microdrilus saliens*, are of small size, an accidental introduction is quite possible, and it is by no means certain that it has not occurred. In any case the genus is known to possess species which are undoubtedly to be reckoned among peregrine forms—such as *D. bolari*, which has turned up in Europe. *Gordiodrilus* and also *Ocnerodrilus* with its sub-genera have very much the same range as has *Dichogaster*. It is to be noted however that these forms are circumtropical, and that their distribution is thus less continuous than that of *Notiodrilus*; they do not however show the markedly discontinuous range of certain other genera of Megascolecidae. For instance *Octochaetus* is well known from New Zealand, and, not occurring in the intermediate tracts, is again met with in India. *Hoplochaetella* is believed by Michaelsen to present us with another precisely similar instance. Then also the genera *Woodwardia* and *Notoscolex* are to be found in Australia and again (absent from the immense tract lying in between) in Ceylon. *Megascolex* has much the same range, showing also this marked and remarkable discontinuity. Stranger still, perhaps, is the range of *Plutellus* and *Megascolides*, of which the former, chiefly found in Australia and Tasmania, not only extends its habitat to Ceylon but also to North America; it is there represented by Eisen's species *Argilophilus marmoratus*, referred by him, and not unnaturally, to a distinct genus, but placed by Michaelsen in *Plutellus*. *Megascolides* is Australian and from the North Island of New Zealand, where its species were regarded by Benham as of a distinct genus, *Tokea*. There is also one form, *Megascolides americanus*, in the western region of North America.

The two genera *Yagansia* and *Chilota*, closely related to *Notiodrilus*, have a range which is short of that of *Notiodrilus*, and we shall see later that there are reasons for regarding these genera as derived from *Notiodrilus*. They are met with only in the south of South America, and in the Cape of Good Hope region.

The range of *Microscolex* seems to be much the same as that of *Notiodrilus*; but it is a little uncertain how far the genus is really autochthonous in the countries where it occurs; and in any case it differs from *Notiodrilus* in occurring in Europe, where the species has been named for a long time *M. phosphoreus*. We do not positively know whether this is 'peregrine' in Europe or not.

The range of the antarctic Acanthodrilinae is in a sense continuous; for they argue the former northward extension of the antarctic continent and in any case they occupy neighbouring land masses. In *Octochaetus* and *Plutellus* the case is different and one of real discontinuity. There are however cases of wide range which is also actually continuous and such is afforded by the genus *Pheretima*. This genus appears to be possibly indigenous to Australia; in any case it reaches from the Solomon Islands on the east to India towards the west, being found in all intermediate continents, while it reaches Japan on the north side of this large area.

There are other genera which extend their range over a considerable area, but which are not so widely distributed as these which we have just been considering. Thus *Diporochaeta* is chiefly Australian but also reaches even the South Island of New Zealand and the southward lying antarctic islands. *Desmogaster* and *Eupolygaster* among the Moniligastridae range from Burmah in the east to Sumatra and Borneo further east, though they are not recorded from intermediate islands. *Perionyx* is found in Burmah, India, Zanzibar, Sumatra, and Java. There are other examples of genera which have much the same range as those enumerated. Finally there are those which are confined to one land mass and very often to a restricted region of that. Thus *Kynotus* is confined to Madagascar, all the genera of Eudrilidae to tropical Africa, some of them, *e.g.* *Beddardiella* and *Euscolex*, to very limited tracts, others to wider or less wide areas in that continent. *Maoridrilus* is only found in New Zealand, to the South Island of which also is confined the genus *Neodrilus*. To the Cape region of Africa is limited *Microchaetus*, and to a belt running across the northern part of the tropical region and extending down the Nile, the remarkable, partly aquatic, *Alma*.

As a kind of appendix to these facts and conclusions we shall next deal with certain widely spread forms that have been already referred to, with the range of different genera over great land masses of the world, and with the earthworms of oceanic islands.

CHAPTER VI

PEREGRINE FORMS

Dr Michaelsen has used this term to describe those species which possess some powers of migration over the sea, denied to the majority of worms, and probably due to the direct interference of man. Thus we find in collections of earthworms from various parts of the world not only examples of forms which do not come from other parts of the world, but also a few which occur in many or even most of such collections. It is for example to be actually expected that a collection of earthworms made in South America, the Philippine Islands, or Australia will contain examples of the apparently ubiquitous *Pontoscolex corethrurus*. This is what has actually happened in cases of which I have personal knowledge, as well as in many others recorded in the literature of the subject. I have myself received this worm from the three parts of the world mentioned, and also from Hawaii. Others have increased its known range to other parts of the South American continent, to Central America, the West Indies, the islands of Sumatra, Java, Borneo, Celebes, Mauritius, and Madagascar, etc. It is in fact found everywhere in the tropics. With this range may be contrasted that of another genus of the same family (Geoscolecidae), viz. *Kynotus*, which, though consisting of many species, is not found outside of the Madagascar district. It should be added that *Pontoscolex* does not appear to contain more than two species, the one not mentioned in the above survey of its distribution being *P. insignis* of Kinberg, which is apparently the same as *P. liljeborgi* of Eisen, and is limited to certain parts of America.

Before attempting to grapple with the remarkable facts implied by the distribution of this genus, it will be well to survey the whole group of Oligochaeta and to reduce to as short a space as possible the total series of facts which are of the same nature.

A case, even more striking than that of *Pontoscolex*, is afforded by the Eudrilid genus *Eudrilus*. As with *Pontoscolex* there are two species of this genus, one, *E. pallidus*, being confined to West Africa, the remaining one, *E. eugeniae*, being world-wide in range. This latter species has received the following names, viz. *E. decipiens*, *E. lacazii*, *E. peregrinus*, *E. sylvicola*, *E. boyeri*, *E. jullieni*, *E. erudiens*, and *E. roseus*; they appear to be all synonyms of the name originally given by Kinberg who however did not recognise the distinctness of the form as a genus. It is now known as *Eudrilus eugeniae*. The variety of names given to supposed different forms (for two of which I am myself responsible) is due to the fact that in earlier days when nothing was known about the geographical distribution of this group of animals it was thought by no means unreasonable that a given genus represented by several species should range over the globe. This fact coupled with imperfect description of structural details led to the multiplication of supposed species, a position which is no longer tenable. This worm is quite as abundant in gatherings from all parts of the world as is *Pontoscolex corethrurus*; and in addition to the countries inhabited by the latter, *Eudrilus eugeniae* has been met with in New Caledonia: tropical Africa is probably its original home.

The two families that have been hitherto considered offer no further instances, among their many species, of worms with so wide a range as those just dealt with. There are indeed one or two forms, *e.g. Criodrilus* and *Glyphidrilus*, which have a considerable range though not nearly equalling that of *Eudrilus* and *Pontoscolex*. These are, however, aquatic forms and the range of aquatic forms is determined as far as we can see by a different series of causes to that of terrestrial forms, which are referred to later.

Among the Moniligastridae we have apparently an instance of a peregrine form. The genus itself has its headquarters in Ceylon and extends a little way in other eastern regions; there is, however, one species, *Moniligaster bahamenis*, described some years since from the Bahamas which must surely be an example of a peregrine form, particularly since it is probably identical with *M. japonicus* whose name is indicative of its habitat.

Among the huge family of the Megascolecidae there are a considerable number of species which apparently possess the same facilities for making their way in different directions and across seas from the locality that is thought to be their real home.

Of the very many genera, however, of which this family is composed, a comparatively small number are thus peregrine in habit at times. All the species known which are distributed broadcast, more or less, over the tropics belong to the genera *Pheretima*, *Microscolex*, *Dichogaster*, *Megascolex*, *Perionyx*, *Ocnerodrilus*, *Kerria*. These several genera are placed in order of frequency of exotic occurrence. Indeed of the two latter genera their frequent life in fresh water may really remove them from the present category altogether. In addition to these are some perhaps more questionable instances, such as the genus *Gordiodrilus* which, prevalently West African, has also been found in the West Indies, in East Africa, and in India and Madagascar. These instances I propose to leave out of consideration in the present sketch. The most obviously peregrine genus of all those enumerated is *Pheretima*, which according to my experience turns up in almost all gatherings of earthworms from any part of the tropical or even sometimes temperate regions of the world. It seems to be fairly well settled that this extensive genus has its real home in the islands of the Eastern Archipelago, perhaps extending a little in various directions from that centre. But examples of the genus have been found in almost all other regions. And what is especially to the point in considering the facts, as will be pointed out with more emphasis later, the assumedly peregrine species do not differ from those found in the real district in which the genus is indigenous.

Dr Cognetti de Martiis enumerates in the Neotropical region, that is in South and Central America and the West Indies, the following species: *Pheretima biserialis*, *P. californica*, *P. capensis*, *P. elongata*, *P. hawayana*, *P. hesperidum*, *P. heterochaeta*, *P. houlleti*, *P. posthuma*, *P. rodericensis*, *P. schmardae* and *P. violacea*. Of these twelve species it is quite certain that the last six occur in the East, where they are doubtless indigenous. So too do the species *P. biserialis*, *P. capensis*, and *P. hawayana*. The synonymy of the different species of this large genus is not yet in a completely settled condition. But in the meantime it is in my opinion quite possible that both *P. hesperidum* and *P. californica* are identical with species also occurring in the East.

There remains the somewhat doubtful *P. elongata* from Peru which has not been very fully described. There is thus no convincing evidence of species really indigenous to and confined to any part of America. Some of these species also occur in many other parts of the world. For instance, *P. heterochaeta* is very widely spread indeed, occurring as it does in Australia, New Caledonia, Madagascar, and even England (in hothouses). This species indeed is the most prevalent of all Pheretimas and seems to be abundant in gatherings of earthworms from various localities as are *Eudrilus* and *Pontoscolex*.

From the island of Madagascar and neighbouring islands the following species of *Pheretima* have been obtained and identified by Dr Michaelsen: viz. *Pheretima pentacystis*, *P. peregrinus*, *P. heterochaeta*, *P. biserialis*, *P. rodericensis*, *P. houlleti*, *P. robusta*, *P. mauritiana*, *P. taprobanae*, and *P. voeltzkovi*. It will be noticed that the majority of these are also included in the list from South America, and that many of them are also found in other parts of the world, and nearly all of them in the East. There remain a few which are doubtful. It is quite possible that *P. mauritiana* is the same as *P. hawayana* and *P. bermudensis*, in which case it has a world-wide range. *P. taprobanae* is well known as a Ceylon species. *P. robusta* also occurs in the East Indian islands. There remain *P. pentacystis*, *P. peregrinus*, and *P. voeltzkovi*.

P. peregrinus is known from Australia and also from Sumatra, so that that species need not concern us in enumerating those which are possibly endemic. In fact it is only *P. pentacystis* and *P. voeltzkovi* which may be really Mascarene.

Another peregrine genus belonging to the subfamily Acanthodrilinae is *Microscolex*. But the limits of this genus may be regarded as at present rather uncertain. And this difficulty somewhat affects the bearing of the facts to be related, though it hardly affects the value of the facts themselves. Dr Michaelsen referred to the genus in his great work seven well-defined species, and four others not so plainly distinct. Of these eleven, two are confined to New Zealand, four to North and Central America, one to Hawaii, one to Madeira, one to Algeria, while the remaining two are found pretty well over the whole surface of the world. More recently the same authority has somewhat extended his view of the generic characters, so as to include a number of forms found in Patagonia, Cape of Good Hope, and the antarctic region generally, while he has lumped together into two species only, viz. *M. phosphoreus* and *M. dubius*, the eleven forms just mentioned, which species therefore are absolutely world-wide in range, and thus form an excellent example of a peregrine form. These species moreover differ from *Pontoscolex* and some others in that they have been able to establish themselves in Europe. Dr Michaelsen also relates that in the cultivated lands of South West Australia, *Microscolex dubius* and *Helodrilus caliginosus* are actually the commonest species; and he calculates that they form together quite 90 % of the earthworms gathered in any locality belonging to this region.

Some of the other Megascolecid peregrine forms will be referred to later. There is no doubt that the family Lumbricidae offers by far the greatest number of peregrine forms and that these are most abundant in collections from extra-European countries, where the collector has searched in cultivated lands. There are at least eight or nine species which are common in many parts of the world though their original home is undoubtedly Europe.

This is a brief review of the facts, more detailed in some cases than in others. It remains to review and compare the results arrived at.

The first general statement that may be made is that this faculty of extending their range beyond the limits assigned by nature is not confined to any one family. For all the chief sub-divisions of the terrestrial Oligochaeta seem to possess it, though in unequal degrees. But the inequality may be more apparent than real. For if it be recollected that the species of the family Megascolecidae are very much more numerous than those of the Eudrilidae or even the

Geoscolecidae, the fact that there are more peregrine Megascolecidae will lose some of its importance. With the Lumbricidae the case seems to me to be different. Here the preponderance, not only in species (relatively speaking) but in individuals, is much above that of other families. This preponderance I should be disposed to assign to the newness of the family coupled with the vigour seen in new races That this is a possible explanation is borne out by the fact that the 'Perichaetidae' (*i.e.* the genus *Pheretima*) is the most salient race of peregrine Megascolecidae, and it is now generally held that this group is the most modern of that enormous family.

Another general statement may be made with even more confidence, viz. that it appears to be an undoubted fact that some species are more capable of extending themselves than others. Thus *Eudrilus eugeniae* occurs everywhere on the great land masses of the globe, except in Europe; it is in fact circummundane in the tropical zone, as is also *Pontoscolex*. *Dichogaster bolavi* is again a trifle more restricted in its range, having been recorded from tropical Africa, South America, West Indies, Madagascar, and India. Its occurrence near Hamburg in Europe is also to be noted. A little more restricted still is *Nematogenia panamaensis* whose range is in Central America, tropical West Africa, and Ceylon. Lastly there are cases such as *Pheretima taprobanae* which, a native of Ceylon, is also found in Madagascar.

It may be asserted in the third place that there are no peculiarities of structure shared by all of these peregrine forms which might account for their physiological similarity, except indeed the somewhat negative feature which they have in common, that is of being of small or moderate size. *Eudrilus* and *Pontoscolex* are not isolated types in their respective families; nor do they seem to approach each other in any respect. Nor can it fairly be said that these peregrine species are marked by any great variability of structure as compared with other forms, which might allow for their suiting themselves to various climates and conditions. It is true that *Eudrilus eugeniae* has received many names which might at first argue some variability. But these names have been perhaps given by persons rather under the influence of the idea that remote habitat implied specific difference, and who were thus inclined to see minute differences, and who perhaps were furthermore led astray in the matter by imperfectly accurate descriptions on the part of others. Certainly some of the peregrine species of *Pheretima* are subject to some variation, particularly in the number and arrangement of their genital pupillae. But this feature is by no means confined to those species and cannot be utilised as in any way an adaptation to wide distribution.

But while we can lay down no general explanation of the phenomenon, it is possible to furnish some explanation of particular cases. Thus the genus *Microscolex* is the only exotic genus which appears to have established itself in Europe, from which country indeed it was early known as an apparently indigenous inhabitant. We must put this and some similar cases down to ability to do without great heat. It is probable in fact that the original home of *Microscolex* is the antarctic half of the globe; and this of itself would allow of its establishing a new home in the northern hemisphere, did other circumstances allow of it.

It might be urged that this genus has been able to establish itself in Europe because it has in fact had the chance denied to other species. There are a good many, however, which would in that case be in the same category. Some years ago I received from time to time a very large number of earthworms from the Royal Gardens at Kew which had been accidentally imported thither from many quarters of the globe, among which I described some eighteen or twenty new species including, for instance, the African genus *Gordiodrilus*. There are plenty of facts of a similar nature and Dr Michaelsen has pointed out that botanical gardens act as centres of dispersion for accidentally introduced Oligochaeta. We must therefore come to the conclusion that temperature is at least one of the causes of a difference in the capability of extending their range shown by the Oligochaeta, a cause which doubtless operates as a check upon extension of range in non-peregrine forms also, and prevents for instance the dispersion of the tropical African Eudrilidae into the region of the Cape.

We may, as it appears to me, confidently look upon indifference to varying temperature as a condition of ability to colonise new countries. But it is obvious that this is not of itself a sufficient cause to explain the facts. Otherwise this country and N. Europe would contain many antarctic earthworms; the only one that has been recorded to my knowledge is *Microscolex*.

Though an inability to endure a temperate climate may have rendered the movements of peregrine species more limited, the same or rather the exactly opposite cause does not seem to have played any important part in this direction. For it is above all the Lumbricidae, normally dwellers in temperate climates, that are so remarkable for their wide range over the world. Nor can it be convincingly asserted that the extra-Palaearctic Lumbricids are real indigenes of those—often tropical—countries. For if so we should expect them to be at least of different species. Lumbricids however from South America, Australia, etc., are specifically identical with European forms.

There is no doubt that wherever land has been at all long under cultivation in any part of the world that land will be found to produce species of the European genera *Lumbricus*, *Helodrilus*, *Eisenia*, etc. More than this the recently imported European forms will be found to have largely or almost entirely driven out the native species, which have retired more into the interior of the country. There is thus here no barrier placed by temperature. It should be remarked, however, that while these earthworms are most abundant in the less tropical regions, they occur in such tropical districts as Peru, though in less striking numbers. Whether those of North America are really indigenes or not remains perhaps a matter for discussion; but it is at least noteworthy that the vast preponderance of species occurring there are also European and even British. In this particular case, which is on the whole the most emphatic of all the cases of peregrine earthworms, some explanations are possible, or at least have been offered. In the first place it would appear that earthworms are more abundant as individuals in northern countries where the soil is rarely dry for prolonged periods. And as has been already pointed out there is a close relation between earthworms and agriculture. Dunghills are fertile gathering grounds for some species, and ploughed fields and gardens always swarm with several species. In the tropics these animals are not so evident; and the strong rays of the sun appear to drive them further underground and into marshes; this obviously lessens the chance of their accidental transference by man. Furthermore Dr Eisen has pointed out that the European species are apt to have clitella and to be fertile all the year round, which is not always the case with other genera. That naturalist has added to this observation the fact that in rich cultivated soils in California it is impossible to find anything but imported European species, since cultivation itself appears actually to drive away the native forms.

CHAPTER VII

THE EARTHWORMS OF OCEANIC ISLANDS

OCEANIC islands are islands that have always been islands, a definition that seems tautological until we compare it with some other land masses that may be termed 'islands.' Geology teaches us in fact that from the point of view of their origin islands may be divided into two quite sharply contrasted classes. There are those detached land masses usually lying near to or comparatively near to some continent, which have been in the course of time detached by the action of the waves from that continent, such as for instance the British Isles, which undoubtedly represent a portion of the European continent which was once quite continuous with Europe. On the other hand we have the Hawaiian archipelago, St Helena, Fernando Noronha, and other similar islands, which are more remote in their position from continents and concerning which it seems clear that they have originated *de novo* by the action of submarine volcanos or of the growth of coral, combined with subsidence, following elevation, or from several of the causes combined. In any case the islands which are termed oceanic islands have never formed part of a continent. They are not relics of previously existing continents. It becomes a matter of great interest to compare the earthworms which are to be found upon oceanic islands with those which inhabit continental islands. Fortunately there are a good many facts at our disposal for this purpose; and we shall compare the earthworms of the Hawaiian archipelago with those which are found upon certain small islands lying to the south of New Zealand, viz. Campbell and Auckland islands and the more southern Macquarie islands.

The earthworms of the Hawaiian archipelago have been studied by a good many persons, and altogether a number of species have been described from that group of islands of which the following is a list: *Pheretima hawayana, P. heterochaeta. P. peregrina,*

P. schmardae, P. hesperidum, P. morrisi, P. perkinsi, P. biserialis (= *P. elongata*), *Allolobophora putris* (= Kinberg's *Hypogaeon havaicum*), *A. foetida, A. caliginosa, A. nordenskioldi, A. limicola, A. rosea,* and finally the well-known *Pontoscolex corethrurus.* Of these species there is only one which is even possibly a form limited to the Sandwich Islands, and that is *Pheretima perkinsi,* a species which I myself at first described as a new form, but which was afterwards regarded as identical with *P. heterochaeta* by Michaelsen, and later still resuscitated by Ude. All the others are found in many parts of the world and not only in the nearest mainland to the archipelago which we are now considering. I have had already occasion to speak of some of them as peregrine forms, especially of *Pontoscolex corethrurus* which occurs all over the world.

The conditions which have been recently revealed by an exploration of the antarctic islands mentioned above are totally different. Dr Benham has enumerated the following species from those islands, viz. *Notiodrilus haplocystis, N. fallax, N. aucklandicus, N. campbellianus, N. macquariensis, Plagiochaeta plunketi, Rhododrilus cockayni, Leptodrilus leptomerus, L. magneticus, Plutellus aucklandicus, Diporochaeta heterochaeta, D. brachysoma, D. helophila, D. perionychopsis* among the Megascolecidae, besides *Phreodrilus campbellianus, Pelodrilus tuberculatus,*

P. aucklandicus and the Lumbricid *Helodrilus constrictus.* There were also four species of purely aquatic Oligochaeta which we shall leave aside from the present enumeration, as their range in space is a matter requiring a different explanation from that of the terrestrial forms. Here we have a series of worms, all of which, save the widely spread Lumbricid, are apparently absolutely indigenous to the islands mentioned since they are all different as *species* from those found elsewhere. Indeed there is a whole genus *Leptodrilus,* consisting, it is true, of but two species, which is a native of the Campbell and Auckland islands and of those only. The other genera are found in the antarctic region, while *Pelodrilus* is still more widely spread.

These facts as will be observed contrast about as strongly as they can with those supplied by the fauna of Honolulu and its adjacent islands. Not only are the worms of the antarctic islands different species from those found elsewhere, but the majority of them do not consist of widely ranging peregrine forms. It appears therefore most probable that these islands are not oceanic islands but a portion of the former existing northern portion of the antarctic continent. Were the species *identical* with those of New Zealand this conclusion would have of course to be reconsidered. The barriers to migration (see chap. VIII) explain the contrast recorded in the foregoing pages.

CHAPTER VIII

MOVEMENT AND MIGRATION AMONG EARTHWORMS

THAT earthworms can move upon the surface of the ground at a rapid pace is probably well enough known to everyone, and that they can also burrow with considerable celerity. Multiplying the inches of progress in minutes of time by centuries with the resulting miles, it is quite clear that there is no reason to suppose that an individual earthworm might not enormously extend its range under favourable circumstances. Their powers of locomotion are such that they could in the course of comparatively few centuries people a continent. As a matter of fact these animals are frequently very widely spread upon a given land surface; but on the other hand they are sometimes equally limited. It behoves us therefore to enquire the reasons for the possibility of extended migration and the causes which have led to its restriction. We are now, it must be borne in mind, considering these animals as purely terrestrial animals moving over the surface of the land by their own unaided efforts. We leave out of consideration any possible assistance in crossing water, whether fresh or salt. We have to consider in fact in the present section the earthworm inhabitants of larger and smaller tracts of continuous land such as the African continent, which will serve as an excellent example wherewith to test the facts and inferences.

And as a 'control' we can compare this continent with the very different continent of Europe.

As an excellent instance, because of the certitude of specific and in most cases of generic distinctions, we may take the Eudrilidae as illustrative of the facts that are to be considered in the present section. That family consists, as will be remembered, of 33 genera at most, which have the following more exact range on the African continent. The genus *Eudriloides* occurs in British and German East Africa and has been met with as far south as Mosambique and even Durban, in which latter locality it has been thought that it is really an accidentally introduced stranger. *Platydrilus* is limited to eastern equatorial Africa, thus not having quite the range of *Eudriloides.*

The small genera (that is small in numbers of species) *Reithrodrilus, Bogertia, Megachaetina, Metadrilus, Notykus* have the same limitation of range as the last genus. *Metschaina* has a wider range from tropical North East to lake Tanganyika. *Stuhlmannia* has a wider range still being found as it is in the Tanganyika district, in tropical North East Africa, and in British and German East Africa near the coast. *Pareudrilus* reaches still further north while *Nemertodrilus* is limited to the Mosambique region and to the Orange River district further south. The only remaining genus of this sub-family of the Eudrilidae is *Libyodrilus* which is purely West African and equatorial.

Of the remaining genera which are usually grouped together into a second sub-family, five, viz. *Malodrilus, Kaffania, Gardullaria, Teleudrilus* and *Teleutoreutus*, are confined to tropical North East Africa. *Eminoscolex* occurs in the same district but also to the south in the great lake region. The most remarkable fact about this genus is that one species *E. steindachneri* comes from the Cameroons, and another *E. congicus* from the Congo, and thus the range of the genus is right across the continent. *Neumanniella* has much the same range. *Polytoreutus* is a purely equatorial East and Central genus, reaching from the coast to the lakes. *Bettonia* known by three species is from British East Africa.

The remaining genera, viz. *Hyperiodrilus, Heliodrilus, Alvania, Iridodrilus, Rosadrilus, Euscolex, Parascolex, Preussiella, Buttneriodrilus, Beddardiella, Metascolex*, are all West African and the vast majority equatorial. We thus see that with one exception the genera of East Africa are totally different from those of West Africa and that the family as a whole is restricted in its range to a comparatively small part of the vast African continent. It also obviously follows, and it is advisable to state this fact however obvious, that no species are common to the two sides of the continent except indeed the ubiquitous *Eudrilus*, whose range over the world has been more than once referred to in this book.

On the other hand the genus *Dichogaster* offers quite different facts, which are in contradiction to those just enumerated. This genus as already said is very characteristic of tropical Africa, and a large preponderance of the known species are confined to that continent. Although there is some variation in structural characters among the many species which compose this genus, there is but little doubt that they are all rightly referred to one genus with perhaps some doubtful, though not very striking, exceptions. In any case the utmost divergence of structure between worms usually placed together in this genus is nowhere near to that which separates the genera of Eudrilidae from each other. Of the African members of the genus the species are pretty evenly divided between the eastern and western halves of the continent; they are, like the Eudrilidae, tropical in range, not occurring to the southward, where their place is taken by the Acanthodrilinae and Geoscolecidae. There are it is true a few species, such as *D. gracilis* and *D. bolavi*, which are common to the two sides of Africa; but in these cases we clearly have to do with those rather mysterious species which can apparently unduly extend their range and which are known as peregrine forms; for they also occur in other parts of the world besides Africa. We have therefore in *Dichogaster* the case of a genus which ranges all over the tropical parts of Africa, but whose species are not common to the Atlantic and Indian shores of that continent.

We will now contrast these conditions, which exemplify certain facts shown by the characteristic Oligochaeta of tropical Africa, with those which obtain in Europe. In this region of the world the prevalent and practically the only genera which need be taken into consideration in surveying the Oligochaetous fauna from the present point of view, are *Lumbricus* and the genus *Allolobophora* of Eisen which has been variously rearranged into genera and sub-genera known by the names of *Helodrilus*, *Bimastos*, *Octolasium*, etc. The structural differences which divide these genera and sub-genera are not great; in any case they do not exhibit such a wide range of variation from each other as do two such Eudrilid genera as *Stuhlmannia* and *Hyperiodrilus*. We find the genera mentioned not only in Europe but extending themselves over more or less of Asia, even occurring in Japan; while the North American continent contains also representatives of the same. Not only do we find this community of genera over vast extents of country greater in diameter than the African continent, but there are also many species which range as widely or nearly as widely as the case may be as the genus to which they belong. Thus the species of *Allolobophora* (we do not trouble about the newer sub-divisions as they hardly affect the facts to be emphasised), *A. caliginosa, A. longa, A. rubida, A. chlorotica, A. octaedra, A. constricta, A. beddardi, Lumbricus terrestris, L. castaneus,* have an enormously wide range over what is generally termed the Palaearctic region, extending also in some cases into the Nearctic. It is true no doubt that the majority, indeed perhaps all, of these are, like certain species of *Dichogaster* mentioned above, among those forms termed peregrine which have the capability of living in every quarter of the globe to which they have apparently been conveyed by man. But there remain many species which have a very extended habitat in the northern hemisphere, and in any case the genera and the species are there truly indigenous and widely spread.

It would thus appear that the capability for independent migration varies greatly among earthworms. Of the types selected for consideration the Eudrilidae are the slowest movers; the genus *Dichogaster* comes next, while the power of migration possessed by the genera *Allolobophora* and *Lumbricus* is very much greater. Assuming for the moment the correctness of this inference it is clear that it will influence many other propositions connected with the relative age of the families of these worms and with many problems of geographical distribution. It appears to us that this simple explanation is the correct one. But to show this it will be necessary to eliminate other possible explanations. It might be urged that the wider range of the genus *Dichogaster* and the still wider range of the genus *Allolobophora* (shown by community of species in widely distant localities) was evidence merely of relative age, that the older groups have had more time to travel and that the newer groups have not had so long a time to spread themselves over their habitat. On this hypothesis the genera of Eudrilidae would be geologically much newer than the genus *Dichogaster* and similar statements might be made for the other forms here under consideration. As already explained we cannot attempt to answer this question in the only way in which it can be really satisfactorily answered, by a reference to fossil forms; for there are no fossils to refer to. So far as comparative anatomy enables us to arrive towards a solution of the question, it would appear that the genus *Dichogaster* belongs to a more ancient race than either of the other two groups considered, and that of these latter the Lumbricidae are the most modern. Moreover we associate not only a wide, but also a discontinuous, distribution with an archaic race; and for this reason also we should place the genus *Dichogaster* in the position of being the most ancient of these Oligochaeta. For the genus occurs in Central America and in certain parts of the East as well as in Africa. So that we can fairly dismiss the view that the Lumbricids by virtue of their greater range over a given area are the most ancient type and that their range is associated merely with their antiquity. Nor does it appear that geographical or meteorological consideration can have had effect in the present instances. For conditions favourable to earthworms prevail in tropical Africa, as in Europe and much of North Asia.

Climate as Affecting Migration.

That excessively rigorous climatic conditions affect the range of earthworms as well as fresh-water forms is quite clear from the conditions which obtain in the most northern climes. At any rate in those regions where physical conditions render it impossible for these Annelids to have their being. A perpetual mantle of snow and a temperature far below freezing point are absolute barriers to the extension of range. And yet there are some few Oligochaeta which do not in the least suffer from a somewhat milder taste of such conditions. Thus species of Enchytraeidae have been met with on glaciers and even found in frozen water, while a few earthworms have been brought from the island of Kolguev. These however are quite exceptions to the general sterility as regards earthworms of the excessively cold regions. We have already seen that there are no general facts to be deduced as concerning the relative abundance of terrestrial worms in the tropics and in more temperate climes. Tropical Africa is, it is true, rich in genera and species; but on the other hand tropical East Indies have but few genera inhabiting their numerous islands. Temperate England has very few genera and not a large number of species; temperate New Zealand has a considerable number of different indigenous genera. When however we leave this general aspect of the question and consider separate families and genera, there seems to be some little relation between climate and distribution and thus some effect of climate in acting as a barrier to migration. For example, though continuity of land surface permits of the tropical African Eudrilidae ranging southwards as far as the Cape they are not met with so far as we know in the most southern parts of Africa; nor are the South American Geoscolecidae found in Patagonia or northward beyond Central America. These instances do really look like an influence of climate upon range. On the other hand we must be careful to eliminate the possibility of another explanation and that is the impossibility of successful migration owing to the previous occupation of the ground with abundant other forms. The very same countries would appear to show that this explanation is unnecessary. For the prevalent genus of the southern tracts of South America *Notiodrilus* extends its way northward as does the same genus from temperate to tropical Africa and Madagascar.

It looks very much, therefore, as if certain Oligochaeta are dependent upon climate for their range, and as if others were at least more independent of climatic conditions. And there are other facts which support this view. The same opinion is supported by the phenomena of involuntary migration, a subject which has been considered also separately under the head of 'Peregrine forms.' The great prevalence of Lumbricidae accidentally imported into many parts of the world shows that temperature is no real bar to their voluntary migration. On the other hand the fact that specimens of the East Indian genus *Pheretima* though commonly imported accidentally into the warmer regions of the world have not been able to make good a footing in Europe, save in greenhouses, shows that this genus is affected in its range by questions of climate. These facts suggest another inference of great interest which can only be mentioned tentatively, and not put forward as a demonstrated conclusion. Seeing that *Lumbricus* (*sensu lato*) can comfortably take up its home in warm extra-European countries, but yet that it has evidently not spread to those countries in the course of nature but by man's interference, it seems possible that time alone has prevented this; and that therefore this family Lumbricidae is one of the most recently evolved families of Oligochaeta. Certain structural features support this way of looking at the matter. The same arguments precisely apply to the genus *Pheretima*, which is also regarded by most systematists as a recently developed race of earthworms. Anyhow the conclusion which the facts seem to warrant is that the effects of climate in influencing distribution are seen to have an unequal effect upon earthworms, some genera being debarred by climatic conditions while others are indifferent to the same.

Mountain Ranges and the Migration of Earthworms.

In many groups of animals the interposition of a lofty chain of mountains presents an insuperable barrier to migration. The barrier is effective for more than one reason. Lack of vegetation and a differing climate are among the more obvious causes which render Alpine chains important as affecting distribution. There is plenty of evidence in the way of positive fact that mountains are not necessarily barriers to the spread of earthworms. The recent explorations of the Ruwenzori chain of mountains in Africa have resulted in the collection of a considerable number of species, some of which come from great altitudes (*e.g.* 4000 metres and slightly upwards), and one species, viz. *Dichogaster duwonica*, which Dr Cognetti de Martiis described from the foot of the glacier Elena. I have in my temporary possession a number of examples of the eastern genus *Pheretima*, some of which are new species from lofty areas in the Philippine Islands. There are plenty of other examples pointing to a like conclusion. It is noteworthy that these forms which have been met with at lofty heights are not essentially different from the plain living forms. One cannot exactly speak, at any rate in the present state of our knowledge, of anything like an Alpine fauna.

It is in fact clear enough that whatever may prove to be the case with regard to particular species, a mountain range is not necessarily a barrier to the dispersal of generic types.

THE OCEAN AS A BARRIER TO MIGRATION.

It is very possible that further investigations into the Oligochaeta will prove that there are more marine forms than those which are enumerated in another chapter. Particularly is this likely to be the case among the family Tubificidae and Naididae. For up to the present those forms belonging to those families which are known to be positively marine in their habit show no great difference from allies inhabiting fresh water, and are in one case indeed (*Paranais*) common to fresh brackish and saline waters. As to earthworms, the number is also extremely limited, and *Pontodrilus* is up to the present the only genus which is known to inhabit a marine situation almost exclusively. It has, moreover, been shown that both earthworms and their cocoons are susceptible to salt water and are killed thereby. Thus the facilities which these animals possess of crossing tracts of ocean are limited by this fact alone, besides other impediments offered by tracts of water as such. We may in fact entirely discount the possibility of earthworms floating across arms of the sea—of any extent at any rate. For they do not swim or float, but sink in water. Possibly when the alimentary tract was entirely empty of earth the worms might float; but it is always full and even if evacuated during their passage to the bottom waters the body thus freed would hardly rise. However the noxious qualities of sea water to earthworms is a sufficient barrier to their traversing even narrow straits. On the other hand it might be suggested that torn up trees especially with the roots and clinging earth still attached might harbour worms and thus transmit them to foreign shores. It has been suggested that in this or in some similar way the species of *Notiodrilus* have been wafted from shore to shore of those lands which are washed by the Antarctic Ocean. Dr Benham, however, in criticising this, calls attention to the violent gales and disturbances of the ocean surface which are so prevalent in those stormy regions, and doubts much whether these animals could retain a safe hold upon some travelling tree trunk. Moreover it is only in this antarctic region where the earthworm fauna of the various continents and islands are so very similar.

Facilities of Migration.

The above brief account of physical features which affect the range in space of the terrestrial Oligochaeta seem to show that the only really important barrier is the ocean; and even a narrow tract of sea water would, as it appears, act fatally in preventing the successful immigration of a race inhabiting one shore to the opposite shore. On the other hand we do undoubtedly find in different countries—even when separated by a large expanse of ocean—closely related forms. The most striking instance of this is that afforded by a consideration of the antarctic species of *Notiodrilus* and *Chilota*. Can this interchange of Oligochaetous faunas be explained by any means which earthworms possess of crossing tracts of sea by the aid of living carriers such as birds? It has been definitely shown that these creatures actually do convey such small animals as Mollusca attached to their feet. Is anything of the kind likely in the case of earthworms? In the first place it may be safely asserted that if it be possible it has not been actually proved. This however might be perhaps put down to the lack of sufficient observation of actual birds and the contents of such masses of soil as are found attached to their feet. A consideration of the habits of earthworms seems to imply that such a mode of transference from country to country is unlikely. In the first place we remark that the general behaviour of earthworms renders this unlikely. Even the smaller kinds, whose bulk would allow of their being carried, are too active in their habits to permit of a safe transference. When disturbed they wriggle and progress with activity. It is not conceivable that they would remain quiescent for sufficient time to allow of a long voyage. But while the bodily transference of adult earthworms seems highly improbable it is conceivable at the first view that their cocoons might be so transferred. We require to know rather more about the cocoons of earthworms before we can accept this view as a possibility; as far as our present knowledge goes it is not likely that these animals can be assisted to emigrate in this way.

For the cocoons are rather bulky for this kind of

porterage. Moreover they are apt to be deposited rather deep down and among the roots of grasses, and in situations where they are not so likely to become entangled in the feet of drinking birds. Assuming, however, that these difficulties can be got over there remains another difficulty. A single cocoon among the terrestrial Oligochaeta does not contain a large number of embryos, as has been pointed out on a previous page. It is true that *Allolobophora foetida* has six within one cocoon, but most of our indigenous forms have but from one to three embryos in a single cocoon. Thus, if successfully imported, it is hardly likely that the developed embryos scattered after their emergence would come together for breeding purposes; and in cocoons with but one embryo the accidental importation in this way would have to be very frequent to produce any result.

The case here is exactly the reverse of that afforded by the aquatic families (or many of them). In these Annelids the attachment of the cocoon to water plants, which are liable to be entangled in the feet of shore-frequenting birds, would tend to favour migration. And in addition to this the cocoons are naturally smaller and often contain a considerable number of embryos. We are to note that the aquatic forms are on the whole distinctly wider in their range than are the earthworms.

CHAPTER IX

THE GEOGRAPHICAL DISTRIBUTION OF EARTHWORMS

THE facts referred to and considered in the last chapter lead to further observations upon the geographical distribution of this group of animals and suggest problems for solution.

It is not the place here to give a general sketch of the division of Biology termed Zoogeography; but a few general conclusions must be laid before the reader in order to render what follows intelligible. It is universally agreed that the range in space (and in time also) of a given species of animal (or plant) is as much a part of its scientific definition as are its anatomical characters. A description for instance of *Acanthodrilus ungulatus* is incomplete without a reference to the fact that it occurs in, and is confined to, the island of New Caledonia.

Each continent or island or part of a continent and part of an island has its own peculiar inhabitants as well as some others which range beyond its confines. Thus as we have seen the genus *Hyperiodrilus* is confined to the tropical West of Africa while the genus *Dichogaster* also found in that region is also met with in other parts of Africa as well as in certain parts of America and of the East. In this way the entire globe may be mapped out into regions characterised by their inhabitants and these regions may also be further subdivided. The commonly accepted regions were originally devised by Mr Sclater and are known as the Palaearctic, Nearctic, Neotropic, Ethiopian, Oriental (Mr Sclater's name was 'Indian'), and Australian. These regions were originally formed to convey the facts relative to the distribution of Passerine birds only; but it is generally held that they apply also to the distribution of vertebrates generally. The science of zoogeography does not however end with the display of maps conveying graphically the mere facts of distribution of this group and that. Its business is also to enquire into the causes of the affinities between the faunas of different regions or the varying degree of remoteness which those faunas may show. On the one hand the varying powers of dispersal and the means of extending their range possessed by different animals have to be considered, and on the other hand geological changes in the relative position of land masses have to be taken into account.

The specific identity between the earthworms of Great Britain and the adjacent part of the continent of Europe would be very difficult to understand were we only acquainted with the fact that salt water is fatal to these animals. But we also know from geology that it was only at a very recent date that England was cut off from union with the continent. Thus an identity of fauna was to be expected. On the other hand we are confronted with a very great difference between the earthworms of eastern tropical Africa and of the adjacent island of Madagascar. In the latter we have as a prevalent form the genus *Kynotus*; in the former continent many Geoscolecidae but no *Kynotus*. It is believed that the separation of Madagascar from the mainland was at an earlier date than that of Great Britain from Europe. We must however be cautious before slipping into what might seem a case of arguing in a circle. It will however probably not be disputed that Madagascar was severed earlier than England.

We will now attempt to map out the world into a series of regions characterised by their earthworm inhabitants and see how far these regions agree with those rendered necessary by the distribution of some other animals.

We can to begin with accept the Palaearctic region. The region however will be a little different from that usually accepted. For we must probably exclude Japan, whose earthworm fauna contains the characteristically Eastern genus *Pheretima*. Otherwise we have a region characterised by the family Lumbricidae, which is really limited to it, and by just a few traces of other genera such as *Hormogaster* among the Geoscolecidae and *Sparganophilus* which however is possibly an accidental immigrant. This region is certainly quite clear. Now according to some persons such as Prof. Heilprin the northern part of America should be joined with Europe and Asia to form an Holarctic region; while by most authors, the separate name of Nearctic is given to the north of the New World. With regard to the terrestrial Oligochaeta it appears to me that this part of the world is possibly to be excluded altogether as possessing no indigenous worms.

In considering the distribution of the Mammalia Sir Ray Lankester excluded New Zealand from his view as never having possessed any indigenous mammalian fauna, and termed this part of the world Atheriogaea. In the same way it is possible that the northern part of the United States and Canada, whose earthworm fauna consists of species of Lumbricidae identical with those of Europe, may possibly be also a region to be excluded in the present survey and spoken of as 'Ascolecogaea.' In the southern part of the United States we shall find genera which will be considered presently. On the other hand it is equally conceivable that this part of the world lost its earthworm fauna through excessive glaciation in the ice age, the forms having been driven south and are now only gradually making their way northwards again. In this case the modern earthworm population which appears to be absent from large tracts of Canada will be simply due to involuntary migration. These two views must be left for further development.

In any case the southern parts of the United States seem to be separable as a distinct region from South America and to be characterised by the sub-family Diplocardiinae, the genus *Diplocardia* extending as far northwards as the state of Illinois. The distinctness of such a region however from Central America and the West Indies is marred by the abundance of *Ocnerodrilus* of which Dr Eisen has described so many forms. On the other hand the West Indies are closely allied in their earthworm fauna to tropical South America, sharing with that region several forms of Geoscolecids belonging in both cases invariably to the sub-family Geoscolecinae. The bulk of the latter are undoubtedly tropical South American in range and there is no doubt whatever about the distinctness of this part of the world as a separate region. There is moreover a further puzzle which confronts us who are trying to delimit an American region or regions. In North America are species of the genus *Argilophilus* which is referred by Michaelsen to the genus *Plutellus* which comes from the East and at least one species of *Megascolides*, also an Eastern genus.

There is at present no doubt to be thrown upon

the indigeneity of *Plutellus*. The species according to Dr Eisen show every sign of being genuine inhabitants of California and like certain New Zealand species such as the *Tokea esculenta* of Benham (referred by Michaelsen to the genus *Megascolides*) were eaten by the natives. If these genera were forms restricted to North America, that is not only with reference to the rest of America but to the world generally, there would be as I think no doubt about the practicability of making a Nearctic region. As it is, it seems to me to suit the facts of distribution better to regard the whole of the land under consideration as forming one great Neogaean region with three sub-regions, the North American, Central American and West Indian, and tropical South American. This region however will not as I take it include the southernmost extremity of South America. Here in Patagonia and in neighbouring islands we have a different earthworm fauna. It is in fact characterised by the sub-family Acanthodrilinae of which it is true some members of the genus *Notiodrilus* extend further north. I shall however defer this part of the subject until the more easy delimitations of regions are disposed of.

Tropical Africa is evidently to be included in a third region which will be defined by the Eudrilidae, Microchaetinae among the Geoscolecidae, and by the great prevalence of *Dichogaster*, a genus whose occurrence in other parts of the tropics is perhaps not yet explained satisfactorily. Also we may record as characteristic of this Ethiopian region a few peculiar genera such as *Nannodrilus* and *Gordiodrilus*. *Alma* being a partly aquatic genus is perhaps less distinctive and as a matter of fact it strays into the Palaearctic region, being found in the lower waters of the Nile. It will be observed that with this exception the limits of the Ethiopian region according to earthworms agrees with that delimitation afforded by a consideration of other groups since it stops short at the Sahara, leaving northern Africa to be referred to the Palaearctic region. At the same time we have an analogy with South America as concerns the southern extremity of the African continent; here we meet with *Notiodrilus* and allied Acanthodrilinae just as in Patagonia and—as also in that quarter of the world—these forms just stray into the Ethiopian region above—specimens of *Notiodrilus* being met with in Madagascar as well as in tropical Africa This bit of Africa as it appears to me must also be cut off from the Ethiopian region and included in an Antarctic region. Madagascar offers a further problem. Are we to include this in Ethiopia or speak of a Malagasy region? Apart from a few forms which are at least possibly to be looked upon as accidental immigrants, such as members of the genera *Pheretima* and *Gordiodrilus*, the fauna of Madagascar consists mainly of many species of *Kynotus*. This genus, a member of the sub-family Microchaetinae, of the family Geoscolecidae, affines Madagascar to Ethiopia and leads me to place both in the same region though we may doubtless speak of a Malagasy sub-region.

We have now to consider the eastern region of the world comprising the two regions known generally to zoogeographers as the Oriental and Australian. Taking a large view of the range of sub-families and genera, and endeavouring to make the great regions of the globe more or less equal, it seems difficult to divide further a region which shall include all of this vast territory, and which may therefore be termed Indo-Australian. For we find as characteristic of the entire stretch of country the great majority of the genera of the huge family Megascolecidae. Indeed the largest sub-family of this family, *i.e.* the Megascolecinae, is, save for the mysterious occurrence of the genera *Plutellus* and *Megascolides* in America, absolutely limited to this area. Another sub-family, that of the Octochaetinae, is limited to it. So far as concerns the others of the sub-families of Megascolecidae it is only the Trigastrinae which occur here (the genus *Eudichogaster* and a few possibly introduced species of *Dichogaster*) and a scattered species or two of *Notiodrilus* of the sub-family Acanthodrilinae. Again there are a few and probably introduced species of the sub-family Ocnerodrilinae. More important still this region has confined to itself the family Moniligastridae; for a species described some years ago by myself from the Bahamas is doubtless an introduced form. We have a complete absence of indigenous Lumbricidae and Geoscolecidae excepting the aquatic *Glyphidrilus* of the sub-family Microchaetinae. It is true that by taking isolated tracts, even large tracts, of this great regional expanse a sub-division into well characterised regions can be apparently recognised. But in taking such a step we shall be confronted with the curious fact that it is rather neighbouring than widely remote sub-divisions which present the greater differences.

If we compare for example India and New Zealand we find in common such striking genera as *Octochaetus*, *Hoplochaetella* and *Diporochaeta*; whereas these genera are absent from the intervening islands of the great Malay archipelago. On the other hand Australia differs from the comparatively neighbouring islands of Borneo and others by the absence in those islands of the characteristic Australian genera such as *Megascolex*, *Notoscolex*, *Plutellus* etc. which are in their turn found in India. It is facts like these which render very difficult the apportioning of the tracts of country forming the eastern hemisphere into separate regions.

There is no doubt that the Malay archipelago and the adjacent coasts of Asia up to Japan differ from both India and Australia by the almost entire limitation of the genus *Pheretima* to them; but we cannot intercalate a region in the middle of another geographical area in this fashion!

The limitations of this great Indo-Australian region now demand consideration. The chief difficulty is offered by the islands of New Zealand and by some of the smaller islands lying far from but still in the neighbourhood of New Zealand. Are we to include New Zealand in this region? There is no doubt that the northern island of New Zealand is much nearer to Australia in its earthworm fauna than is the southern island. There are, it is true, a number of genera peculiar to New Zealand, which are *Rhododrilus, Leptodrilus, Maoridrilus, Neodrilus, Plagiochaeta, Pereiodrilus, Dinodrilus, Dinodriloides,* but these do not represent the whole of any family or even sub-family and they have all of them near relations in other parts of the region as has been pointed out —even to the peninsula of India itself. Again New Zealand contains members of the genus *Notiodrilus,* that characteristic Antarctic form. In fact New Zealand would appear to be a transitional zone between an Indo-Australian and an Antarctic region.

The last region into which the world can be divided according to its fauna of earthworms is an Antarctic. I am of distinct opinion that this region is quite necessary in spite of the views of some others.

Although the genus *Notiodrilus* certainly, and *Microscolex* possibly, extend into the tropical regions of America, Africa, and Australia, these species are but few, and the bulk of the species and of the allied genus *Chilota* are restricted to the antarctic quarter of the globe; they also extend all over it, that is to say in the southernmost parts of South America, in the Cape region of Africa, in Kerguelen and the Crozet Islands, and in New Zealand, as well as in the Auckland Islands and other neighbouring islands. It is true that I have excluded New Zealand from this region on the grounds that it forms a debateable ground between it and the Indo-Australian. But apart from this part of the world the rest of the territories mentioned should be combined to form the antarctic region.

Having therefore arrived at a mapping out of the world into regions in accord with its earthworm fauna, it is desirable to ascertain what light the facts throw upon the geological and evolutionary questions with which the study of zoogeography deals. The existence of an antarctic region binding together such distant points as South Georgia, the Cape of Good Hope and Kerguelen Island, seems to argue strongly for the former extensions northwards of the antarctic continent so far north as to embrace these several regions of that hemisphere. In view of the facts relating to the danger of sea water to earthworms, to their lack of facilities for migration, other than unassisted locomotion, points which have been dealt with earlier, it is difficult to explain their range in the antarctic hemisphere on other grounds. The very fact that the actual earthworm fauna of New Zealand has led us on the whole to assign it to the Indo-Australian regions shows the inherent uselessness of the current view of zoogeography. For were we to leave the matter here the relationship of New Zealand to the regions of the world which lie to the south of it would not be apparent. However, here as in so many cases there is an antagonism between cut and dried systems and the indications of evolution.

This assumed existence of a former antarctic continent which connected Southern Africa and Southern America as well as various islands has perhaps a further justification in the distribution of the Geoscolecidae. This family is divisible into two well-marked sub-families of which one as has already been mentioned is limited to South America and another practically to Africa (the exceptions being species of the largely aquatic *Glyphidrilus*), while a third sub-family the Criodrilinae is more widely distributed—again in accordance, one may perhaps assume, with its largely aquatic mode of life. It is also conceivable that the genus *Dichogaster* is another example pointing the same way. The arguments for regarding this genus as an indigene of the

East are not strong. But there is on the other hand no doubt that the Indian *Eudichogaster* is very closely allied to it. But it is by no means excluded from this argument to suppose that these Trigastrinae owe their likeness to convergence. At any rate there are examples of equally marked convergence which seem to be as nearly proved as can be in another though allied group. The New Zealand *Neodrilus* is to all intents and purposes a *Maoridrilus* in which one of the two pairs of spermiducal glands and spermathecae has disappeared. It retains the characteristic alternation in the position of the nephridia of *Maoridrilus*, and other structural similarities unite the two genera. In the same way species of *Microscolex* seem as easily derivable from *Notiodrilus*. *Microscolex* and *Neodrilus* are so near that had we no such hint of their origin it would be reasonable to place them in the same genus. They at least show a marked convergence.

It will be noticed therefore that the facts of their distribution agree, as it would appear, with the structure of the terrestrial Oligochaeta. The primitive characters of the genus *Notiodrilus* are to be seen in the double spermaries and glands appended to the duct, and the corresponding spermatheca, in the absence, or very slight development, of the papillae, so frequent in more specialised genera such as *Pheretima*, and in the general simplicity of many organs of the body which are more complicated elsewhere. As one would expect with an archaic form this genus is widely ranging, being found in all the principal land masses of the globe except in the Euro-Asiatic continent.

Furthermore geographical facts would at least be not contradictory to the view that this genus, and therefore the terrestrial Oligochaeta generally, originated in the Antarctic hemisphere and that in pushing northwards it has given off various descendants which survive in the various regions of the world. Basing our views of the possibilities of range among earthworms on the actual facts already dealt with, it would seem that the peopling of America from Africa or of Africa from America, if it has occurred, has not taken place through Europe and the north generally. For otherwise we should expect traces of the passage. It is true that we actually have *Hormogaster* as a possible sign that the Geoscolecidae have passed this way. But that is an isolated case and may be referred to the extension northwards of this particular genus rather than as an indication of a whole migration through those territories. Another conclusion which a collocation of the various facts brought together in this book appears to lead to is that the group of the terrestrial Oligochaeta is relatively speaking a modern one.

Convinced as we must be of the fact that range

is only possible by unaided locomotion through continuous land areas, the fact that but few gaps occur in the range of a particular sub-family or lesser group seems to indicate that no great time has elapsed since the specialisation of these different forms. The dependence of earthworms upon vegetable mould also points in the same direction and furnishes an argument for the belief that these animals only greatly increased on the advent of abundant dicotyledonous plants, and perhaps indeed were actually contemporaneous with them.

LIST OF LITERATURE REFERRING TO EARTHWORMS

In the list given below I am only able to mention a few of the larger works relating to this group. To give anything like a complete list would demand many pages of titles. From the works selected the reader can, if it be desired, find his way to the remaining literature of the group.

A. General works

Vejdovsky. System und Morphologie der Oligochaeten. Prag, 1884.

Beddard. A Monograph of the Oligochaeta. Oxford, 1895.

Michaelsen. Oligochaeten in 'Das Thierreich.' Berlin, 1900.

Michaelsen. Die Geographische Verbreitung der Oligochaeten, 1903.

Vaillant. Annelès in Suites à Buffon. Paris, 1886.

B. Earthworms of (1) Australia

Fletcher. A series of papers in Journ. Linn. Soc. New South Wales, 1886–90.

Spencer. A series of papers in Proc. Roy. Soc. Victoria, 1892–5.

Michaelsen. In Die Fauna Sudwest-Australiens. Jena, 1907.

(2) New Zealand and Antarctic Islands

Benham. Report on Oligochaeta of the Subarctic Islands of New Zealand. Wellington, N. Z., 1909.

Benham. A series of papers in Quart. Journ. Micr. Sci. 1904, Proc. Zool. Soc. 1904, 1905 and Trans. N. Z. Inst., 1901–10.

Beddard. In Trans. Roy. Soc. Edinb. 1891 and Proc. Zool. Soc. 1889.

(3) ASIA

Michaelsen. The Oligochaeta of India etc. in Memoirs Indian Mus., 1909.

(4) EUROPE

Rosa. Revisione dei Lumbricidi. Mem. Acc. Torino, 1893

(5) AFRICA

Michaelsen. A series of papers in Mitth. Naturhist. Museum. Hamburg, 1891–1911.
Beddard. Quart. Journ. Micr. Sci., 1890–95.

(6) AMERICA

Eisen. Mem. Calif. Acad., 1894–96
Cognetti de Martiis. Mem. Acc. Torino, 1905–6.
Rosa. Ibid., 1895.
Beddard. In Hamburg. Magalh. Reise, 1895 and Nachtrag to same by Michaelsen.

Also numerous other works by the above-named authors and by Perrier, Horst, Ude, Lankester, Stolc, Pierantoni, Friend, Stephenson, Southern, Goodrich, etc., etc.

INDEX

CAMBRIDGE: PRINTED BY JOHN CLAY, M.A. AT THE UNIVERSITY PRESS

Ingram Content Group UK Ltd.
Milton Keynes UK
UKHW021816130323
418485UK00006B/516